GLASGOW UNIVERSITY PUBLICATIONS
L

THE FABRICIAN TYPES OF INSECTS IN THE HUNTERIAN COLLECTION AT GLASGOW UNIVERSITY

THE FABRICIAN TYPES OF INSECTS IN THE HUNTERIAN COLLECTION AT GLASGOW UNIVERSITY

COLEOPTERA

PART II

By

ROBERT A. STAIG, M.A., Ph.D., F.R.S.E.

LECTURER IN ZOOLOGY (ENTOMOLOGY)
UNIVERSITY OF GLASGOW

CAMBRIDGE
AT THE UNIVERSITY PRESS
1940

CAMBRIDGE UNIVERSITY PRESS
Cambridge, New York, Melbourne, Madrid, Cape Town,
Singapore, São Paulo, Delhi, Mexico City

Cambridge University Press
The Edinburgh Building, Cambridge CB2 8RU, UK

Published in the United States of America by Cambridge University Press, New York

www.cambridge.org
Information on this title: www.cambridge.org/9781107688803

First published 1940
First paperback edition 2013

A catalogue record for this publication is available from the British Library

ISBN 978-1-107-68880-3 Paperback

CONTENTS

ILLUSTRATIONS

INTRODUCTION

This volume is a further contribution towards a descriptive Catalogue of the Insect Types in the Hunterian Collection at Glasgow University.

As stated in the Introduction to Part I, the special interest and importance of Dr William Hunter's Collection of Insects is due to the fact that many of the specimens are the types of insect species founded by J. C. Fabricius, Antoine G. Olivier, Dru Drury and other early entomologists. Therefore it seemed advisable to publish descriptions together with figures of these valuable types and thus make them more accessible for purposes of systematic entomology.

The usefulness of such work has been questioned by one authority, mainly on the ground of defective examples; it has, however, been appreciated by others, notably and in a very understanding way by Dr Walther Horn, Director of the Institute of Entomology, Berlin. He emphasized that 'it is more important to give good descriptions and serviceable illustrations for the future than to continue for ever keeping the historical things sacred and thereby to a great extent "bolting the door" against Science'.

Whatever view may be held as to the value of this publication, it has apparently proved serviceable to specialists, judging from the interest shown in the first published part; it has made known to them the whereabouts of several types and the descriptions of these have helped them to clear up points at issue and settle questions of identity to their satisfaction.

It is a pleasant duty to acknowledge especial indebtedness and express my thanks to the *Carnegie Trustees* for again awarding a special grant towards the cost of production, and to the *Publications Standing Committee of Glasgow University* for similar substantial assistance as formerly to ensure the publication of this second volume.

I also desire to thank Professor Edward Hindle for his promotive approbation of this Catalogue, initiated by his predecessor in the Chair, Emeritus Professor Sir John Graham Kerr, to whom I feel grateful for his continued interest.

The types of Coleoptera herein described (under their *modern* names and in the order adopted in Junk and Schenkling's *Coleopterorum Catalogus*) are those of species belonging to the Families *Endomychidae*, *Coccinellidae*, *Helodidae*, *Buprestidae*, *Elateridae*, *Tenebrionidae*, *Oedemeridae*, *Rhipiphoridae*, *Meloidae* and *Pyrochroidae*; and, as previously mentioned, the descriptions are necessarily limited to those features or characters which I have been able to make out satisfactorily, many of the types being in a more or less defective state with age.

With reference to *Carabus unicolor* Fab., *Carabus pallipes* Fab. and *Scarabaeus fricator* Fab., which are described in Part I, supplementary notes on these are included at the end of this volume. By their helpful interest and special knowledge, Professor Edwin C. van Dyke and Dr Henry C. Fall have enabled me to reach a more definite conclusion as to the identity of the first and second above-mentioned insects. The references and synonyms of several of the types described in Part I have been revised according to the *Coleopterorum Catalogus* (Junk and Schenkling), and these revisions are also included in the supplement.

I have again been fortunate in having the assistance of Miss Margaret Rankin Wilson, D.A. (G.S.A.). Her drawings are admirable portrayals, revealing her sound sense of colour and her skill in rendering the subtle shades of these insects. I am much indebted to her for her constant interest and careful, painstaking work.

ROBERT A. STAIG

THE ZOOLOGY DEPARTMENT
THE UNIVERSITY, GLASGOW
November, 1939

Order COLEOPTERA (*continued*)

Super-family *DIVERSICORNIA* (*continued*)

Family ENDOMYCHIDAE

56. *Aphorista vittata* (Fab.)

Coleopterorum Catalogus, pars 12 (E. Csiki, 1910), Endomychidae, p. 40. *Catalogue of the Coleoptera of America, North of Mexico* (Charles W. Leng, 1920), p. 209.

North America.

SYN. *Tritoma vittata* Fab., *Mant. Ins.* I, p. 44, No. 4 (1787); *Ent. Syst.* I, 2, p. 506, No. 6 (1792).
Catops vittatus Fab., *Syst. Eleuth.* II, p. 564, No. 3 (1801).
Endomychus lineatus Oliv., *Ent.* VI, 100, p. 1072, pl. 1, fig. 2 (1808).
Eumorphus distinctus Say, *J. Acad. Nat. Sci. Philadelphia*, V, 1825, p. 303.

The type of this species is noted in the card-index of Dr Hunter's Collection as missing; but in Cabinet A, drawer 3, there is a specimen, misplaced under label

'Hab. in Anglia',

which answers the descriptions given by Fabricius and Olivier and resembles Olivier's figure of *lineatus*. This insect has been compared with modern examples of the species in the British Museum, and it is evidently the missing type.

Description of Type, *Tritoma vittata* Fab. Form elongate-ovate, the elytra narrowing in front and behind, convex, smooth and shining; the upper surface reddish brown with patches of black, the under parts and the legs light reddish brown with a scanty covering of very short and fine whitish hairs.

The *head* is reddish brown, oblong and insunk in the prothorax; a deep crescentic suture marks off the flattened frons from the convex clypeus. The vertex, frons and clypeus are thinly punctulate, the punctules bearing short yellowish white hairs; the labrum also is punctulate and lightly covered with fine hairs; the mandibles are angulate and the tips are produced as long and pointed blades; the other mouth-parts are imperfect in the specimen. The eyes, which are brownish black, oblong and rather narrow, extend from the vertex to the gula; the facets are large and very convex. The antennae are wanting; the bases of the antennae are situated between and above the eyes and on each side of the frons anteriorly.

The *pronotum* is transverse, narrower than the elytra and plano-convex, the disc being a little raised; the front is deeply hollowed and has a thin translucent stridulatory membrane which fills the distinct notch in the middle of the excavated front margin and which projects slightly beyond it; the sides are bisinuate and contracted behind; the base is rather narrow, nearly straight and broadly margined; the front angles are strongly produced and almost sharp, the hind angles are sharply produced, nearly rectangular. The pronotum is strongly marginate, except at the frontal notch where the margin is extended as the stridulatory membrane.

The surface of the pronotum is finely but not very closely punctured with scattered punctules. Upon the disc, directly in front of two black spots and near the front margin, there are two conspicuous pits transversely placed; the basal channel, the channels extending to the hind angles, and the lateral foveae, which extend forward to the middle of the disc, are deeply impressed.

The colour of the pronotum is reddish brown, except the margin, which is black, and the disc which is infuscate around two black spots roughly reniform in shape and transversely placed.

The *prosternum* is deep, very convex and smooth; the prosternal process, which is narrow and truncate, meets the mesosternum. The *mesosternum* is narrow between the coxae

Tritoma vittata Fab. × 12

and wider anteriorly; it has a median elevation, it is hollowed on each side and it is bordered anteriorly and laterally by a carina. A straight furrow marks the junction of the meso-sternum and the *metasternum*, which has a rounded and raised anterior margin. The posterior margin of the meta-sternum is between and behind the widely separated posterior coxae; it has the form of a distinct and broadly triangular flange cleft in the middle and thus forming two small tri-angular lobes. The posterior marginal groove, which marks off the flange, is sharply angulate. A finely impressed median longitudinal line extends from the apex of the posterior marginal groove about halfway towards the anterior margin. The metasternum is finely punctulate, the punctules bearing short and fine whitish hairs.

The *scutellum* is transverse and broadly rounded behind.

The *elytra* are moderately convex, oval, widest about the middle, with the apices gently rounded and with well-developed epipleura. The outer borders of the elytra, the apices and the inner (sutural) borders as far as the middle of the suture are narrowly marginate. The elytral surfaces are regularly but not closely punctulate, the punctules bearing very short silky whitish hairs; and on the inner side of each shoulder callus there is a roughly triangular impression.

The elytra are reddish brown with an elongate black patch along the middle of each outer margin, and with a similar but longer patch forming an irregular border along each inner (sutural) margin, except the apical portion.

The *legs* are moderately long and stout and are lightly covered with short and fine whitish hairs; the anterior coxae are globose and closely approximated, just separated by the thin prosternal process which projects between them; the intermediate or middle coxae are globose and not closely approximated, the posterior coxae are transverse and widely separated; the coxae and femora are separated by large blade-shaped trochanters; the femora are club-shaped, their distal portions being swollen, the tibiae also are distally enlarged, the anterior and middle tibiae are curved, all the tarsi are wanting.

The groove between the metasternum and the base of the abdomen is deeply marked. There are five visible *abdominal sterna*, which are finely punctulate, and the punctules bear short and fine whitish hairs; the basal abdominal sternite is very large, its length being greater than that of the three succeeding sterna together; the fifth sternite is about half the length of the basal one, and its hind margin is sinuate with a distinct and wide median excision.

Length 5·5 mm.; breadth (across the elytra) 2·5 mm.
Hab. India (Fab.), East Indies (Oliv.).
See Plate 29.

Family Coccinellidae

The following are the species of Coccinellidae mentioned by Fabricius, in his published works, as having been described by him from specimens in Dr Hunter's Collection:

Coccinella glacialis	*Syst. Ent.* p. 80, No. 12 (1775).
Chrysomela 10-*maculata*	*Ibid.* p. 105, No. 60.
Coccinella biguttata	*Mant. Ins.* 1, p. 59, No. 72 (1787).
ursina	*Ibid.* p. 61, No. 98.

The above names are the original names as given by Fabricius, and the references are to the works in which these species were first described.

The following species was described by Olivier from a specimen in Dr Hunter's Collection:

Coccinella annulata[1] *Ent.* VI, 98, p. 996, pl. 2, figs. 19*a* and *b* (1808).

[1] Previously described by Voet, *Cat. Syst. Col.*, *La Haye*, 2, pl. 45, fig. 9 (1766) and by Linnaeus, *Syst. Nat.* ed. XII, p. 579 (1767).

57. *Brachyacantha ursina* (Fab.)

Coleopterorum Catalogus, pars 118 (R. Korschefsky, 1931), Coccinellidae, I, p. 207. *Catalogue of the Coleoptera of America, North of Mexico* (Charles W. Leng, 1920), p. 212.

U.S.A., Mexico.

SYN. *Coccinella ursina* Fab., *Mant. Ins.* I, p. 61, No. 98 (1787); *Ent. Syst.* I, I, p. 291, No. 116 (1792); *Syst. Eleuth.* I, p. 386, No. 157 (1801); Oliv. *Ent.* VI, 98, p. 1054, pl. 2, figs. 14*a* and *b* (1808).
ab. *congruens* Casey, *J. New York Ent. Soc.* VII, 1899, p. 117.
ab. *sonorana* Casey, *Canad. Ent.* XL, 1903, p. 413.

The single specimen in Cabinet A, drawer 6, under label

'*Coccin. ursina*
Fabr. MSS'

is apparently the type; it answers the descriptions given by Fabricius and Olivier, except that the spot markings are not white, as stated by Fabricius, but yellow; it also resembles Olivier's figure and it closely agrees with modern examples of the species in the British Museum.

Description of Type, *Coccinella ursina* Fab. Form broadly oval, moderately convex (subhemispherical), surface glossy.

The deeply insunk *head* is yellow and punctulate; the clypeus has, behind the antennal notch, a short knob-like projection overlying the eye; the front margin of the clypeus is sinuate and marginate. The labrum, reddish yellow and with some golden hairs on its surface, is transverse and the anterior angles are rounded; the distal segment of the maxillary palps is thickly enlarged, oblong, somewhat conical and very obliquely truncated. The eyes, pale green, are round and prominent, with the facets slightly convex. The antennae are reddish yellow, moderately long and inserted in a notch of the clypeus at the front of the eyes, a little on the inner side; the basal segment of the antenna is largely exposed, it is partly covered by a thin and reddish semicircular projection overarching the base of the notch. One antenna is defective, and the other is so placed that it is difficult to get a complete view under the microscope. The number of antennal segments appears to be ten; the first or basal segment and the second are larger than the succeeding five, the third is longer than the fourth, fifth, sixth and seventh; the antennal club is fusiform and is composed of three unequal segments, the apical one being the smallest and the antepenultimate one the largest.

The *pronotum* is transverse and marginate; the arcuate front margin is slightly emarginate in the middle, the anterior and posterior angles are rounded. The pronotum is black with the anterior and lateral portions yellow and the black-yellow boundary line is irregular; the surface of the pronotum is punctulate, and here and there the spaces between the punctules show (under the microscope) an irregular kind of reticulation. The *mesepimera* are yellow and roughly triangular with the apices at the middle coxae. The *thoracic sterna*, and the *coxae*, are brownish black and punctulate, the punctules bearing short and fine hairs.

The *scutellum* is small and has the form of an equilateral triangle; it is black and a little rugulose.

The *elytra*, moderately convex, are at the base almost as broad as the pronotum and they are narrowly marginate except at the base; the humeral callosities are prominent and the apices are rounded. Upon each black elytron there are five large yellow spots (2, 2, 1); the first two are transversely placed at the base, the outer one is triangular and fills the outer or humeral angle, and the inner one, which is roughly hemispherical, is on the base of the elytron and adjoining the scutellum; the third and fourth spots also are transversely placed about the middle of the elytron, the third is subspherical and on the outer margin, the fourth is roughly spherical and near the sutural margin, and each spot has two small projections posteriorly; the fifth spot is roughly spherical and is situated near the apex of the elytron. The horizontal

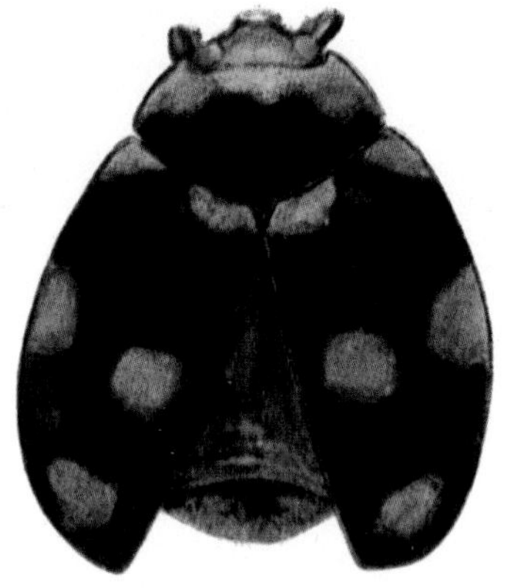

Coccinella ursina Fab. × 18

epipleura are a little wider than the metepisternum and become narrowed in front of the first abdominal sternite. The surface of the elytra, like that of the pronotum, is punctulate with a fine irregular reticulation between the punctules.

The *legs* are punctulate and thinly clothed with fine golden hairs; there is a distinct denticle on the outer margin of the anterior tibiae, about the middle of their length.

The *abdomen* has ventrally six free segments. The posterior border of the fifth abdominal sternum is sinuate, that of the sixth is widely emarginate. The coxal lines of the first abdominal sternum are roughly semicircular, but incomplete towards the outer side. The first sternum is the longest, the second is nearly as long as the first, the third and fourth are about half the length of the first, the fifth is, at the outer sides, as long as the second. The abdominal sterna are glossy black and punctulate, the punctules bearing short fine hairs.

Length 3 mm.; breadth 2·5 mm.
Hab. North America (Fab.).
See Plate 30.

58. *Ceratomegilla* (*Megilla*) *maculata* De Geer

Mém. Hist. Ins. v (De Geer, 1775), p. 392, t. 16, f. 22. *Coleopterorum Catalogus*, pars 120 (R. Korschefsky, 1932), Coccinellidae, II, p. 312. *Catalogue of the Coleoptera of America, North of Mexico* (Charles W. Leng, 1920), p. 215.

U.S.A., California, Canada, Central and South America, West Indies.

SYN. *Chrysomela* 10-*maculata* Fab., *Syst. Ent.* p. 105, No. 60 (1775); *Sp. Ins.* I, p. 131, No. 85 (1781); *Mant. Ins.* I, p. 75, No. 112 (1787).
Coccinella 10-*maculata* Fab., *Sp. Ins.* I, p. 98, No. 29 (1781); *Mant. Ins.* I, p. 57, No. 44 (1787); *Ent. Syst.* I, 1, p. 276, No. 50 (1792); *Syst. Eleuth.* I, p. 367, No. 63 (1801); Oliv. *Ent.* VI, 98, p. 1016, pl. 3, figs. 40*a* and *b* (1808).

The insect in Cabinet A, drawer 4, under label

'*Chrys.* 10-*maculata*
Fabr. pag. 131, No. 85'

is evidently the type. Fabricius in his description of *Coccinella* 10-*maculata* in *Ent. Syst.* refers to his *Chrysomela* 10-*maculata* as 'mera hujus varietas'. Comparing the type with Olivier's figure of *Coccinella* 10-*maculata*, the chief difference appears to be confluence of the pronotal spots and of the apical spots on the elytra of the latter.

Description of Type, *Chrysomela* 10-*maculata* Fab. Form elongate oblong, moderately convex, the surface slightly glossy and punctulate, coloration orange-red with large black spots.

The *head*, deeply insunk in the thorax, is glossy black with a median longitudinal triangular orange-red patch of very regular isosceles form, its base upon the fronto-clypeal line and its apex at the front margin of the pronotum. The surface of the head is punctulate and finely granulate between the punctules. The frons is flattened, slightly depressed. The clypeus, which is orange-red, is short and transverse, it is marked off by an impressed and slightly arcuate line representing the fronto-clypeal suture, and it is narrower than the space between the insertions of the antennae; its sides are excavated where the antennae are inserted and at the outer side of each antennal excavation there is a short knob-like projection overlying the eye. The clypeo-labral suture is well marked and slightly arcuate. The labrum is dark brown, transverse, rounded in front, its width is the same as that of the clypeus, its length is fully one-third of the length of the head; its surface is thinly punctulate and finely granulate between the punctules, and the punctules bear whitish hairs which are longest at the sides. The last segment of the maxillary palps is dark brown, it is very obliquely truncate and has the form of a somewhat flattened cone. The eyes are subspherical and prominent, with the supraorbital ridges

Chrysomela 10-*maculata* Fab. × 15

raised and somewhat prominent in front; the facets are slightly convex and under the microscope the finely faceted surface has a silvery grey appearance with vermiculate black markings. The antennae are defective, almost entirely wanting; the antennal insertions are within excavations of the clypeus at the front of each eye and a little on the inner side, and the basal segments are exposed.

The *pronotum* is orange-red with two large black spots which are roughly trilobed in form, transversely placed and near together; the greater part of the disc is occupied by these spots, which extend from the base of the pronotum and reach nearer to the sides than the front. The pronotum is transverse, broadest across the middle, and a little convex; in front it is slightly and widely emarginate and also slightly sinuate; its sides are rounded, the front angles are a little advanced and rounded, the hind angles are rounded; its base is widely lobate about the middle; it is marginate all round, narrowly on the front and on the base, wider on the sides and translucent. The surface of the pronotum is punctulate and finely granulate between the punctules.

The *mesepimera* are roughly triangular, with the blunted apices at the middle coxae, and partly convex, the convexity being confluent with the convexity of the *mesepisterna*. The mesepimera are black and punctulate, the punctules bearing short and fine whitish hairs.

The *prosternum* is convex, a little arched mid-frontally, and it has a narrow brown margin. The prosternal process is convex and terminally rounded, and it extends between but not beyond the anterior coxae. The colour of the prosternum is orange-red, the surface is thinly punctulate, and the punctules bear fine yellowish hairs.

The *mesosternum* is glossy black. The *metasternum* is convex and abruptly declivous to the wide posterior border, and it is marked with a lightly impressed longitudinal line which begins in the posterior dimple at the border and becomes obsolete towards the front. The surface of the black metasternum is rugulose and punctulate, some of the punctules bearing short and fine whitish hairs. The metasterno-abdominal suture is angular and is situated on a level before the posterior coxae.

Apparently the prothorax has been separated from the rest of the body and glued on; but as the parts are badly fitted together, it is not possible to describe the scutellum and the bases of the elytra.

The *elytra* are orange-red with six large black spots on each elytron arranged 2, 1, 2, 1 at about equal distances apart. The two spots at the base are placed obliquely transverse; the spot on the suture is confluent with the corresponding one on the other elytron, and together they have a pear-shaped form with the narrow end at the base of the elytra; the other spot is on the shoulder callus and is subspherical. The third spot, which is the largest, is situated in the middle of the elytron and is transverse and irregular in form. The fourth and fifth spots are transversely placed and roughly rounded; the fourth, which is on the suture, is largely confluent with the corresponding spot on the other elytron, the fifth is situated near the outer margin. The sixth spot, which is the smallest, is roughly oval; it is situated between the suture and the outer margin and near the apex. The elytra are about one-third wider than the prothorax and moderately convex; the shoulders are moderately prominent; the sides are rounded at the shoulders, very slightly hollowed about the middle third of their length, slightly divergent beyond the middle, where the elytra are widest, and rather sharply rounded in towards the apices, which are separately rounded off. The sides of the elytra are also broadly margined and the margins are strongly punctate; the apices are narrowly marginate. The surface of the elytra is punctulate and finely granulate between the punctules. The *epipleura*, which are horizontal and about twice the width of the metepisternum, become narrowed at the first abdominal sternum and gradually taper to the apices. The surface of the epipleura is punctulate and the punctules bear very short and fine golden hairs.

The *legs* are long, glossy black and thinly covered with short yellowish hairs; the first two segments of the tarsi are thickly golden pubescent beneath.

The *abdomen* has six free ventral segments. The first abdominal sternum is without coxal lines. The fourth sternum is marked by a slightly arched bordering line near the posterior margin; and there is a similar but feebler line on the third and second sterna. The posterior border of the fifth sternum is sinuate and that of the sixth is emarginate. The first sternum is the longest, the second is about half the length of the first, the third and fourth are about equal in length and are shorter than the second, the fifth and sixth are about equal in length and are shorter than the fourth. The abdominal sterna are glossy black with broad orange-red lateral borders; their surface is punctulate and finely granulate between the punctules, which bear very short and fine yellowish hairs.

Length 5·5 mm.; breadth (across the elytra) 2½ mm.
Hab. America (Fab.), South America, Antilles (Oliv.).
See Plate 31.

59. *Hippodamia glacialis* (Fab.)

Coleopterorum Catalogus, pars 120 (R. Korschefsky, 1932), Coccinellidae, II, p. 341. *Catalogue of the Coleoptera of America, North of Mexico* (Charles W. Leng, 1920), p. 215.

Pennsylvania, Missouri, Indiana, Massachussetts, North Carolina.

SYN. *Coccinella glacialis* Fab., *Syst. Ent.* p. 80, No. 12 (1775); *Sp. Ins.* I, p. 96, No. 19 (1781); *Mant. Ins.* I, p. 56, No. 34 (1787); *Ent. Syst.* I, I, p. 274, No. 39 (1792); *Syst. Eleuth.* I, p. 364, No. 50 (1801); Oliv. *Ent.* VI, 98, p. 1007, pl. 5, fig. 68 (1808).

Coccinella abbreviata Fab., *Mant. Ins.* I, p. 54, No. 14 (1787); *Ent. Syst.* I, I, p. 269, No. 19 (1792); *Syst. Eleuth.* I, p. 360, No. 27 (1801); Oliv. *Ent.* VI, 98, p. 1006, pl. 3, fig. 26 (1808).

Coccinella remota Weber, *Observationes Entomologicae*, p. 49, Kiliae, 1801.

The three specimens in Cabinet A, drawer 6, under label

'*Coccin. abbreviata*
Fabr. MSS'

answer the descriptions of *abbreviata* given by Fabricius and Olivier and correspond with Olivier's figure; they also answer the description of *glacialis*, which is essentially the same as that of *abbreviata*. One of the specimens has also been compared with the British Museum modern examples of *H. glacialis* and it closely agrees.

The type of this species, described by Fabricius under the name *glacialis*, was stated by him to be in Drury's Collection; his later description of the same insect, under the name *abbreviata*, is said to be from a specimen in the collection of D. Blagden. Probably, as Professor Graham Kerr suggested,[1] the three Hunterian examples are metatypes, determined by Fabricius as conspecific with the Blagden type; and it is also probable that one of them is the Drury insect on which Fabricius founded *glacialis*.

Description of Metatype, *Coccinella abbreviata* Fab. Form oblong ovate, moderately convex, the surface glossy and punctulate, the head and the pronotum yellow and black, the elytra reddish brown and black, the under parts mainly brownish black.

The deeply insunk *head* is polished black with a fronto-clypeal and central lozenge-shaped yellow-coloured patch, which is continuous on each side with a lesser one of similar shape but transversely placed and extending on the fronto-

[1] 'Remarks upon the Zoological Collection of the University of Glasgow' by Professor J. Graham Kerr (*Glasgow Naturalist*, Vol. II, No. 4, September 1910, p. 103).

orbital processes. The surface of the head is punctulate, the punctules bearing short and fine hairs. The frons is flattened and it has a knob-like process in front of each eye. The transverse clypeus, which is marked off by a feebly impressed arcuate line representing the fronto-clypeal suture, is narrowed towards the front; its sides are excavated at the insertions of the antennae. The clypeo-labral suture is well marked and straight. The labrum is black and transverse; it is slightly emarginate in front, rounded at the sides, and the anterior angles are rounded; its width is that of the front of the clypeus and its length is less than one-third of the length of the head; its surface is punctulate and the punctules bear yellowish hairs which are longest at the sides. The eyes are subspherical and moderately prominent; the facets are convex, and under the microscope the finely faceted surface has a golden appearance with small black spot markings. The antennae, which are reddish brown and moderately long, are inserted in an excavation of the clypeus, in front of the eyes (fronto-orbital process) and on the inner side; the first or basal segment is the largest, the second is about half the length and breadth of the first, and the third segment, which is the same length as the second but not as thick, is longer and thicker than the succeeding five, which are about equal in length and thickness; the antennal club is composed of three segments and is fusiform, but blunted at the apex. The maxillary palps are large and four-segmented; the second segment is oblong, the distal segment is flask-shaped with the tip bluntly rounded off.

The *pronotum* is moderately convex, transverse and marginate; the front is widely excavate and slightly arcuate about the middle, the sides are rounded with rim-like margins, the base is sinuate with a slight and broad median lobe towards the scutellum, the anterior and the posterior angles are rounded. The pronotum is polished black and bordered with yellow along the front and the sides. There is a narrow triangular extension of yellow from the middle of the front border on to the black disc, and upon the black disc there are two yellowish longitudinal vittae femur-shaped and obliquely placed. The pronotal surface is punctulate.

The *mesepimera* are yellow and triangular with the apices at the middle coxae.

The *thoracic sterna* are glossy black and punctulate, the punctules bearing fine yellowish hairs.

The *scutellum* is of triangular form with sinuate sides; it is black and punctulate.

The *elytra*, moderately convex, are, across the base, as broad as the pronotum and broader across the middle third of their length; the bases are sinuate and marginate, the sides are not perfectly rounded and the apices are rounded in at the suture; the margin of the base and of the outer border forms a narrow rim. On the slightly prominent shoulder callus there is a small black spot; a similar but larger black spot, roughly triangular in shape, is situated near the apex; and behind the middle of each elytron there is a large transverse and rounded black spot with an extension outwards and obliquely backwards, thus forming a fascia roughly angular between the suture and the outer border. The three metatypes show variation of the black markings on the elytra; two of the metatypes are without the black spot on the shoulder callus, and on one of these the black transverse fascia near the middle is thinner and the black apical spot much larger. The surface of the elytra is punctulate like that of the pronotum. The horizontal *epipleura* are at the base twice the width of the metepisternum but become gradually narrowed towards the apex.

The metasterno-abdominal suture is angular and is situated on the anterior level of the posterior coxae.

The *coxae* and the *legs* (imperfect in this specimen) are glossy black, punctulate and clothed with fine yellowish hairs.

The *abdomen* has ventrally six free segments; the posterior border of the fifth abdominal sternum is slightly angular,

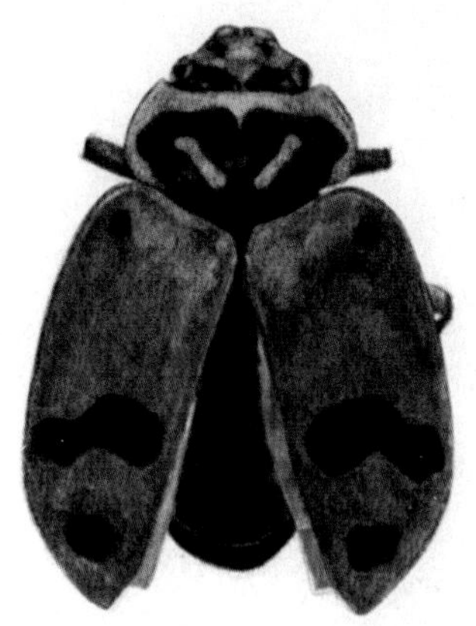

Coccinella abbreviata Fab. × 10

that of the sixth is slightly emarginate. The sterna are glossy black with orange-coloured outer borders, which are slightly marked on the first and sixth, strongly marked on the second to fifth and in the form of angular patches on the second, third and fourth sterna. The surface of the sterna is closely punctulate and lightly set with short yellowish hairs.

Length 6 mm.; breadth (across the elytra) 4 mm.
Hab. North America (Fab. and Oliv.).
See Plate 32.

60. *Adalia bipunctata* var. *annulata* Linn.

Coleopterorum Catalogus, pars 120 (R. Korschefsky, 1932), Coccinellidae, II, p. 388.

Palaearctic Region, North America, North and Central Africa.

SYN. *Coccinella annulata* Voet., *Cat. Syst. Col., La Haye*, 2, pl. 45, fig. 9 (1766); Linn. *Syst. Nat.* ed. XII, p. 579, No. 5 (1767); Fab. *Sp. Ins.* I, p. 94, No. 8 (1781); *Mant. Ins.* I, p. 53, No. 11 (1787); *Ent. Syst.* I, I, p. 268, No. 14 (1792); *Syst. Eleuth.* I, p. 359, No. 20 (1801); Oliv. *Ent.* VI, 98, p. 996, pl. 2, figs. 19*a* and *b* (1808).

The type of this species is pre-Fabrician. The single example in Cabinet A, drawer 6, under label

'*Coccin. annulata*
Fabr. pag. 94, No. 8'

is the insect (from Germany) described and figured by Olivier; and it has been compared with modern examples of the species in the British Museum Collection.

Olivier's coloured figure (19*b*) is a useful representation, but the colour pattern is rather diagrammatic.

The brief description of this var. *annulata* given by Fabricius in his *Species Insectorum* and repeated in his subsequent works is apparently quoted from Linnaeus and is without reference to any particular collection. However,

Fabricius may have determined the Hunterian specimen as being conspecific with the Linnaean type, and therefore to be regarded as a homotype of the species.

Description of Homotype, ***Coccinella annulata*** **Linn.** Form ovate, very convex, the entire surface glossy and closely punctulate; the head black with two yellow patches; the pronotum black with lateral borders of reddish yellow; the elytra orange-red with two sickle-shaped black markings anteriorly, a broad black band across the middle, and posteriorly a narrower black band which bends abruptly forwards at right angles to join the middle band. The under parts of the body finely punctulate, thinly pubescent and glossy black, except the abdominal sterna which are reddish at the sides. The antennae and mouth-parts reddish brown, the legs black.

The insunk *head* is black with two patches of yellow roughly oval and closely adjoining the eyes, the yellow extending on to the small fronto-orbital processes. The surface of the head is finely punctulate. The transverse clypeus is excavate at the sides, it is narrowed a little at the front and is there also slightly excavated; its front angles project a little forwards and outwards and between these small projections the clypeo-labral suture is deeply marked and straight. The sutural border of the clypeus is yellow. The labrum, which is narrower than the clypeus and punctulate, is black proximally; the distal portion is yellow, rounded in and slightly emarginate. The eyes are fairly prominent and the facets are convex; the finely faceted cornea is brown with a golden lustre. The exposed knob-like basal segment of each antenna is inserted, in front of the eye and on the inner side, on the excavated side of the clypeus between the anterior projection and the fronto-orbital process.

The *pronotum* is transverse and it is widely and deeply excavate in front, the anterior angles are on a level with the middle of the eyes. The front of the pronotum is very slightly arcuate between the anterior angles, which are strongly rounded; the disc is convex and the sides are flattened out, forming rims which are markedly extended in front; and as the sides diverge gently outwards to the rounded posterior angles, the pronotum is therefore broader behind than in front. The base is broadly lobate. The front and the side rims of the pronotum are narrowly marginate and its surface is punctulate. The disc is glossy black, the side rims are reddish yellow and there is a faint border of the same hue along the front margin.

The *metepisternum* is closely punctate with rugulose effect.

The thoracic sterna are glossy black and punctulate, the punctules bearing short fine hairs; the *prosternum* is convex and rugulose.

The small *scutellum* is black with the centre reddish, sharply triangular and equilateral.

The *elytra*, very convex and finely marginate except on their sutural sides, are, at the base, broader than the pronotum; the bases are distinctly sinuate and the shoulder angles are strongly rounded; the sides are strongly rounded out about the middle and gradually rounded in towards the apices, which are bluntly pointed and a little divergent, and the humeral callus is prominent. The surface of the elytra is punctulate and the colour pattern, orange-red with very conspicuous black markings, is as follows: A broad transverse black band extends across the middle of the elytra from rim to rim, it is irregular in outline and its thickest portion on each elytron is a distinct bulging just beyond the inner or sutural half. Midway between this band and the apex of the elytra there is a similar transverse band, which is narrower and also shorter; it bends abruptly forwards directly under the bulging of the thicker one and thus becomes confluent with it longitudinally. Two markedly rounded areas of red, side by side at the suture, are thus almost entirely encircled; for the black of the two bands extends slightly along the suture on each side. Above the middle band there are two lesser black markings, roughly sickle-shaped and close together, one on each elytron and extending obliquely outwards from the base at the side of the scutellum; the two 'handles' are contiguous at the bases and divergent at the 'hooks', the concavities of which face inwards and the tips almost reach the middle band.

The reddish yellow horizontal *epipleura* are anteriorly about one and a half times the width of the metepisterna and are gradually narrowed towards the apex.

The *legs* (and coxae) are glossy black and punctulate, with fine yellowish hairs.

The *abdominal sterna* are punctulate and lightly clothed with short and delicate recumbent yellowish hairs. The first three sterna are entirely glossy black; the outer portions of the fourth, fifth and sixth sterna are reddish and the posterior borders of the fifth and sixth are almost straight.

Length 4 mm.; breadth 3 mm.
Hab. Europe in gardens (Fab.), Germany (Oliv.).

61. *Adalia decempunctata* (Linn.)

Syst. Nat. ed. x, 1758, p. 366; *Fn. Suec.* 1761, p. 155; *Syst. Nat.* ed. xii, 1767, p. 581. *Coleopterorum Catalogus*, pars 120 (R. Korschefsky, 1932), Coccinellidae, ii, p. 411.

Europe, Asia, North Africa.

Syn. *Coccinella biguttata* Fab., *Mant. Ins.* i, p. 59, No. 72 (1787); *Ent. Syst.* i, i, p. 284, No. 80 (1792); *Syst. Eleuth.* i, p. 374, No. 100 (1801); Oliv. *Ent.* vi, 98, p. 1033, pl. 2, figs. 9*a* and *b* (1808).

Coccinella variabilis Fab., *Gen. Ins.* p. 218 (1777); *Sp. Ins.* i, p. 104, No. 62 (1781); *Mant. Ins.* i, p. 60, No. 85 (1787); *Ent. Syst.* i, i, p. 287, No. 101 (1792); *Syst. Eleuth.* i, p. 380, No. 130 (1801); Oliv. *Ent.* vi, 98, p. 1046, pl. 7, fig. 105 (1808).

The type of this species is pre-Fabrician; and Korschefsky places *biguttata* Fab. under *decempunctata* Linn. provisionally.

The specimen in Cabinet A, drawer 6, under label

'*Coccin. biguttata*
Fabr. MSS'

is evidently the insect described by Fabricius and Olivier; and the figure of it in Olivier's work is a very good representation. Owing to its frail condition, I have not been able to compare this type with modern examples in the British Museum Collection. It does not resemble any of the varieties of *decempunctata* Linn. in the Bishop Collection; but it closely approaches two examples of a series labelled *variabilis* Ill. in the Bishop (Armitage) Palaearctic Collection.

Description of Type, *Coccinella biguttata* Fab. Form ovate, somewhat oblong, moderately convex, the surface glossy, glabrous above, and punctulate; the head and pronotum black with yellow, the elytra reddish brown with two large suffused light yellow basal markings and with some irregular blackish blotching. The under parts of the body punctulate, lightly hairy and glossy black, except the abdominal sterna which are partly reddish brown. The antennae, mouth-parts and legs reddish brown.

The deeply insunk *head* is black with a broad transverse band of yellow between the eyes; this yellow band is incurved in front (fronto-clypeal suture) and the colour is continuous on the fronto-orbital processes. There is also a narrow transverse strip of yellow, which has a slightly arched hind border, at the clypeo-labral suture. The surface of the head is punctulate, the punctules bearing very short and fine hairs, and on the vertex there is a short row of longer yellowish hairs. The frons is flattened and it has a small fronto-orbital process in front of each eye. The transverse clypeus is narrowed towards the front, its sides being obliquely excavated; the front angles of the clypeus form small anterior lobes projecting upwards and outwards, and between these anterior lobes the clypeo-labral suture is well marked and straight. The labrum is black with a yellowish border; it is transverse, slightly emarginate in front, rounded at the sides, and the anterior angles are rounded; its surface is punctulate, the punctules bearing yellowish hairs which are longest at the sides. The eyes are moderately prominent; the facets are convex and the finely faceted cornea has a golden appearance with black markings. The antennae are inserted in front of the eyes and on the inner side, each in an excavation of the clypeus between its anterior lobe and the fronto-orbital process; the small anterior clypeal lobe does not entirely cover the base of the antenna. The distal segment of the antennal club is somewhat truncate at the apex. The large terminal segment of the maxillary palps is obtuse-angular and obliquely truncate (securiform).

The *pronotum* is moderately convex, transverse, and widely and deeply excavate in front, the anterior angles being nearly on a level with the anterior portions of the eyes; the front is also slightly arcuate between the anterior angles, which are strongly rounded. The sides of the pronotum have narrow marginate rims and diverge gently outwards to the slightly rounded posterior angles between which the pronotum is broader than in front; the base is strongly and broadly lobate and inturned towards the scutellum. The front margin of the pronotum is narrowly bordered with yellow, the disc is polished black, and the sides are broadly bordered with yellow; the inner boundaries of the yellow side borders are sinuate. The pronotal surface is punctulate.

The *metepisterna* are closely punctate with rugulose effect.

The thoracic sterna are glossy black; the *prosternum* is convex and rugulose; the *mesosternum* is punctate; the *metasternum* is punctulate, the punctules bearing short fine hairs, and it is also marked by two small foveae (horizontally placed) on each side and by a median longitudinal furrow which arises at the metasterno-abdominal suture and becomes obsolete anteriorly.

The *scutellum* is black, sharply triangular and equilateral, and punctulate.

PLATE 33

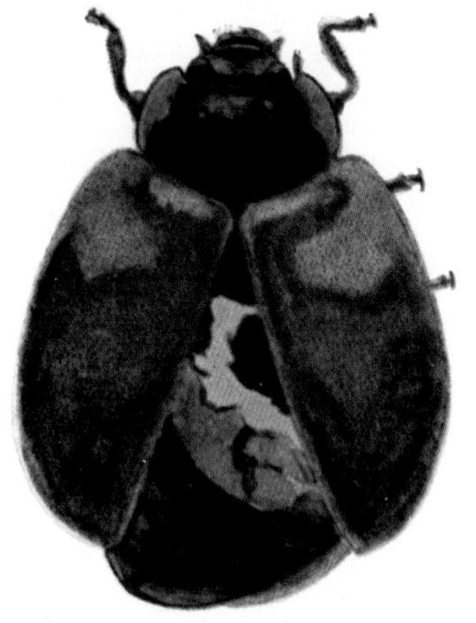

Coccinella biguttata Fab. × 15

The *elytra*, moderately convex and finely marginate all round, are, across the base, as broad as the pronotum and a little broader across the middle third of their length; the bases are slightly sinuate and the shoulder angles are strongly rounded; the sides are subparallel about the middle and gradually rounded in towards the apices, which are bluntly pointed, and the humeral callus is moderately produced. The elytra are punctulate and reddish brown, with wavy streaks and blotches of black here and there throughout; and on each elytron there is a distinct broad and transverse light yellow fascia, which extends from the outer margin about the humeral angle and round the humeral callus posteriorly where it bends abruptly upwards and reaches about half-way across towards the suture. The outline of this fascia is irregular and obscure. The reddish brown, horizontal *epipleura* are, at the base, nearly twice the width of the metepisterna and become gradually narrowed towards the apex.

The metasterno-abdominal suture is angular and is situated on the anterior level of the posterior coxae.

The *legs* (and coxae) are reddish brown, punctulate and set with fine yellowish hairs.

The segments of the *abdomen* are more or less thinly punctulate and lightly clothed with short fine hairs. The coxal lines of the first abdominal sternum are semicircular and complete; this sternite and also the second and third are glossy black with reddish brown outer borders; the fourth, fifth and sixth sterna are entirely reddish brown, and the posterior borders of the fifth and sixth sterna are straight.

Length 4½ mm.; breadth 4 mm.
Hab. Europe (Fab. and Oliv.).
See Plate 33.

Family DERMESTIDAE

62. *Dermestes felinus* Fab.

Coleopterorum Catalogus, pars 33 (K. W. von Dalla Torre, 1911), Dermestidae, etc., p. 43.

Australia.

SYN. *Dermestes felinus* Fab., *Mant. Ins.* I, p. 34, No. 11 (1787); *Ent. Syst.* I, 1, p. 229, No. 12 (1792); *Syst. Eleuth.* I, p. 314, No. 13 (1801).
Dermestes ater De Geer, *Mém. Hist. Ins.* IV, p. 223, t. 18, f. 7 (1774); *Abh. Gesch. Ins.* IV, p. 130, t. 18, f. 7 (1781); Oliv. *Ent.* II, 9, p. 9, pl. 2, figs. 12*a* and *b* (1790).
Dermestes piceus Thunb., *Nov. Ins. Spec.* I, p. 8 (1781).

As stated by Professor Graham Kerr,[1] the type of this species is said (*Mant. Ins.* I) to be in the Banks Collection and the two specimens in Dr Hunter's Cabinet A, drawer 3, under label

'*Der. felinus*
Fabr. MSS'

are possibly metatypes. One of the specimens has been compared with the Banksian type in the British Museum and it closely agrees.

Family HELODIDAE

Sub-family HELODINAE

63. *Elodes minuta* Linn.

Syst. Nat. ed. XII, p. 645 (1767). *Coleopterorum Catalogus*, pars 58 (M. Pic, 1914), Helodidae, etc., p. 24.

Europe.

SYN. *Cistela pallida* Fab., *Syst. Ent.* p. 117, No. 8 (1775); *Sp. Ins.* I, p. 148, No. 11 (1781); *Mant. Ins.* I, p. 86, No. 17 (1787); *Ent. Syst.* I, 2, p. 46, No. 26 (1792); Oliv. *Ent.* III, 54, p. 7, pl. 1, figs. 10*a* and b (1795).

[1] 'Remarks upon the Zoological Collection of the University of Glasgow' by Professor J. Graham Kerr (*Glasgow Naturalist*, Vol. II, No. 4, September 1910, p. 103).

Cyphon pallidus Fab., *Syst. Eleuth.* 1, p. 501, No. 1 (1801).
Galleruca melanura Fab., *Ent. Syst.* 1, 2, p. 22, No. 45 (1792).
Cyphon melanurus Fab., *Syst. Eleuth.* 1, p. 502, No. 6 (1801).

There are two examples of this species under label

'*Cistela pallida*
Fabr. pag. 148, No. 11'

in Cabinet A, drawer 3. Both specimens are in a very imperfect state, one being without the right elytron and the greater part of the body, antennae, legs, etc. They correspond with the descriptions given by Fabricius and Olivier; and I have been able to establish their identity by comparing them with a British Museum modern example of *Elodes minuta*, with which they closely agree.

Description of Co-type, *Cistela pallida* Fab. Form oblong-ovate (the sides rather parallel) and moderately convex; coloration mainly fulvous, but fuscous on the disc of the thorax and pitchy around the apices of the elytra, and with the eyes, the margin of the thorax, the margins and suture of the elytra, and the abdomen apparently black. Clothed with a light pubescence of fine, recumbent, pale yellow hairs.

The *head* is small; it is, at the base, about one-half of the breadth of the thorax; it is narrowed towards the front and its sides are strongly excavated; the surface is punctate with irregular punctures. The clypeo-labral suture is slightly arcuate; the labrum is transverse and finely punctate, the punctules bearing fine golden hairs. The eyes are large; the antennae (both imperfect) are yellowish and are inserted in front of the eyes.

The shape of the *thorax* (*pronotum*) is semicircular; the middle part of the base is broadly lobate towards the scutellum and the scutellar portion of this lobe is distinctly arcuate; the posterior angles are rounded, it is marginate all round, and its surface is punctulate and lightly pubescent.

The *scutellum* is large and of a triangular shape; but the

middle of the base is produced as a slight and arcuate lobe, the sides are a little incurved and the apex is rounded. The surface of the scutellum is punctulate and lightly pubescent.

The *elytra* are a little broader than the thorax, elongate, narrow, broadest at the middle and there subparallel; the shoulders are prominent and with rounded humeral angles, the apical or sutural angles are rounded, and the sutural margins are slightly divergent towards the apices; the outer and the inner or sutural borders are marginate. The elytral surface is closely punctulate and finely pubescent, and it is marked with some faint longitudinal linear ridges.

The *epipleura* are narrow, fulvous and punctulate, the punctules bearing light yellowish hairs.

The under parts of the body (*thoracic and abdominal sterna*) are badly damaged and only the upper parts of two legs and a fragment of the abdomen now remain. The femora are fulvous, punctulate and hairy.

Length 5 mm.; breadth (across the elytra) 2½ mm.
Hab. England (Fab. and Oliv.).
See Plate 34.

Family CANTHARIDAE

64. *Cantharis viridescens* Fab.

Mant. Ins. 1, p. 167, No. 3 (1787); *Ent. Syst.* 1, 2, p. 214, No. 5 (1792); *Syst. Eleuth.* 1, p. 295, No. 6 (1801).

The type of this species is, as stated by Fabricius, in the Banks Collection. The two specimens in Cabinet B, drawer 6, under label

'*Canth. viridescens*
Fabr. MSS'

agree with the type and with modern examples in the British Museum Collection.

Cistela pallida Fab. × 25

Family LAMPYRIDAE

Sub-family LUCIOLINAE

65. *Luciola cincta* (Fab.)

Coleopterorum Catalogus, pars 9 (E. Olivier, 1910), Lampyridae, p. 41.

Poulo Candore.

SYN. *Lampyris cincta* Fab., *Mant. Ins.* I, p. 161, No. 6 (1787); *Ent. Syst.* I, 2, p. 99, No. 7 (1792); *Syst. Eleuth.* II, p. 101, No. 9 (1801); Oliv. *Ent.* II, 28, p. 23, pl. 3, fig. 22 (1790).

The single example of this species in Cabinet B, drawer 6, under label

'*Lamp. cincta*
Fabr. MSS'

answers the descriptions given by Fabricius and Olivier. The type is in the Banks Collection, British Museum.

Family CLERIDAE

66. *Trichodes irkutensis* Laxmann

Nov. Comment. Acad. Petrop. XIV, I, 1770, p. 595, t. 24, f. 4. *Coleopterorum Catalogus*, pars 23 (S. Schenkling, 1910), Cleridae, p. 90.

Siberia, Mongolia, Alps, Istria, Galicia.

SYN. *Clerus bifasciatus* Fab., *Sp. Ins.* I, p. 202, No. 7 (1781); *Mant. Ins.* I, p. 126, No. 11 (1787); *Ent. Syst.* I, 1, p. 208, No. 11 (1792); Oliv. *Ent.* IV, 76, p. 9, pl. 1, fig. 9 (1795).
Trichodes bifasciatus Fab., *Syst. Eleuth.* I, p. 283, No. 3 (1801).

This species is represented by one specimen in Cabinet B, drawer 9, under label

'*Clerus bifasciatus*
Fabr. MSS'

I have compared this example with the type of *bifasciatus* Fab., which is in the Banks Collection (British Museum), and they closely agree.

Family Buprestidae

The following are the species of Buprestidae mentioned by Fabricius, in his published works, as having been described by him from specimens in Dr Hunter's Collection:

Buprestis punctatissima	*Syst. Ent.* p. 217, No. 5 (1775).
blanda	*Sp. Ins.* I, p. 276, No. 23 (1781).
fasciata	*Mant. Ins.* I, p. 177, No. 13 (1787).
cyanipes	*Ibid.* p. 178, No. 30.
aurata	*Ibid.* p. 178, No. 33.
3-punctata	*Ibid.* p. 179, No. 34.
dorsata	*Ibid.* p. 179, No. 38.
meditabunda	*Ibid.* p. 183, No. 80.
ruficollis	*Ibid.* p. 184, No. 85.

The above names are the original names as given by Fabricius, and the references are to the works in which these species were first described.

The following species was described by Olivier from a specimen in Dr Hunter's Collection:

Buprestis quadrimaculata *Ent.* II, 32, p. 76, pl. 10, fig. 110 (1790).

67. *Polycesta cyanipes* (Fab.)

Coleopterorum Catalogus, pars 84 (J. Obenberger, 1926), Buprestidae, I, p. 43. *Monographie des Buprestides* (C. Kerremans), I, 1906, p. 474, pl. 9, fig. 5. *Catalogus Buprestidarum* (Ed. Saunders, 1871), p. 59. *Annals and Magazine of Natural History*, Vol. XIV, Seventh Series, 1904 (Chas. O. Waterhouse), p. 254.

Jamaica.

Syn. *Buprestis cyanipes* Fab., *Mant. Ins.* I, p. 178, No. 30 (1787); *Ent. Syst.* I, 2, p. 196, No. 46 (1792); *Syst. Eleuth.* II, p. 196, No. 55 (1801); Oliv. *Ent.* II, 32, p. 39, pl. 9, fig. 104 (1790).

Polycesta resplendens Thoms., *Typi Bupr. Mus. Thoms.* 1878, p. 45.

P. jamaicensis White.

Two specimens in Cabinet B, drawer 8, under label

'*Bup. cyanipes*
Fabr. MSS'

are the co-types of this species, they have been compared with the modern examples of *cyanipes* Fab. in the British Museum Collection.

These co-types were examined by the late Charles O. Waterhouse;[1] he stated that *Polycesta resplendens* Thomson is clearly the same insect, also that the smaller of the two specimens is that figured by Olivier. I find, however, on comparing the two co-types with Olivier's illustration and description, that it was the *larger* one which he figured and described; this is apparent from the outline of the front part of the pronotum and from the coloration (thorax blue in the middle, green at the sides; elytra green, bluish about the suture) mentioned in his description.

Description of Co-type, *Buprestis cyanipes* Fab. This, the larger example, is a *female*. Form elongate-oblong, subparallel, flattened and somewhat insunk on the pronotum and around the scutellar area, a little angular in outline at the sides of the prothorax and slightly sinuate along the sides of the elytra, which are gradually narrowed and serrate towards the apices; the prothorax, at its widest part, and the elytra are equal in width; the dorsal surface is punctate and rugulose on the head and prothorax, closely punctate-striate on the elytra and also rugulose on the sutural half of each elytron; the striations have a more or less notched appearance. Coloration bright violet above with some metallic green about the sides, mainly metallic green beneath, the legs and tarsi cyaneous. The surface of the under parts is irregu-

[1] 'Remarks upon the Zoological Collection of the University of Glasgow' by Professor J. Graham Kerr (*Glasgow Naturalist*, Vol. II, No. 4, September 1910, p. 106). 'Observations on Coleoptera of the Family Buprestidae, with descriptions of new species' by Charles O. Waterhouse, F.E.S. (*Annals and Magazine of Natural History*, Vol. XIV, Seventh Series, 1904, p. 254).

larly punctate, the puncturation is coarse and close, presenting a honeycombed appearance, on the thoracic parts and particularly on the pro-episterna and the prosternum.

The *head* is violet-coloured and narrow, it measures one-half the breadth of the thorax; it is hollowed in front and its surface is rugose-punctate, the punctures bearing short yellowish hairs. The vertex is convex and broad, and is marked by a finely impressed median longitudinal line. The frons is hollowed before the clypeus and between the antennae, and the coarse puncturation gives this fronto-clypeal hollow a honeycombed appearance. At each side of the fronto-clypeal hollow there is an obliquely placed crescentic ridge which partly encloses the hollowed antennal area in front of each eye. The clypeus is short, broad and arcuately emarginate in front in the middle, around the base of the labrum, which is damaged in this example. The labrum of the smaller co-type is long, it is punctate and emarginate with a thick fringe of long yellowish hairs. The gena is closely punctuate and the mentum and gula are rugose-punctate. The elliptical and obliquely placed eyes, rich reddish brown with irregular dark spots and bordered with black, are narrow, not very prominent and not close together. The antennae are inserted in the rounded hollows in front of the eyes and above the anterior angles of the clypeus; the first and longest antennal segment is club-shaped, the second and shortest is knob-like, the third and fourth are longer than the fifth, which is slightly longer than the six successive segments. The last six segments are slightly serrate. The antennal segments are dark metallic green tinged with blue.

The *prothorax* is transverse, its breadth is fully twice its length, it is broadest behind the middle where the sides become angular, and it is obliquely narrowed towards the front; the narrow front is margined by a thickened ring which is interrupted at the prothoracic episterna.

The *pronotum* is deeply excavate and sinuate in front; its base is sinuate and has a smooth border marked off from the

Buprestis cyanipes Fab. ♀ × 4

raised disc, the middle part of the border forming a small scutellar lobe. The base of the disc is marked by a median longitudinal cleft. The sides of the pronotum (viewed from above) are roughly rounded or curved; but the curvature is distinctly angulate at a short distance from the posterior angles, which are right-angled. The side margins have the form of irregular carinae which are inflected and hidden from view above, except at the posterior angles, and which are obsolete towards the forwardly projecting anterior angles. The pronotum is not evenly convex, for the disc is flattened about the middle and slightly depressed. The pronotal surface is slightly rugulose and is irregularly punctate; the punctures are large and close together on the slopes of the disc, smaller and more apart on the flattened area towards the base.

The sides or pleura (*prothoracic episterna*) are broadly triangular, a little hollowed and rugose-punctate, the punctures bearing short recumbent yellowish hairs.

The *prosternum* is flattened about the median line and is irregularly punctate; the punctures, which bear fine yellowish hairs, are close together on the sides and become diminished both in size and number on the prosternal process towards its apex. The prosternal process (between the anterior coxae and reaching to the metasternum) is flattened and it is thinly and irregularly punctate, the oval punctures bearing golden hairs.

The *mesosternum* and the *metasternum* are punctate; the punctures, which bear fine golden yellow hairs and which are smaller than those on the prosternum, are closer together (with rugulose effect) on the side portions and on the epimera and episterna, and are fine and thinly scattered on the middle portions of the sterna. The metasternum has an interrupted median groove and is emarginate. The ante-coxal portion of the metasternum is distinctly marked off by an arcuate raised line, the ends of which reach the base; and it is divided by a median longitudinal sulcus, which is broadly hollowed out in the middle at the point where it crosses the arched ante-coxal line. Each half of this ante-coxal arch is rendered conspicuous by a narrow and cyaneous coloured depression, closely adjoining the middle portion in front and beset with small punctures close together.

The *scutellum* is a small smooth boss with two excavate sides.

The *elytra* are gently convex, except about the base. The length of the elytra is two and a half times the width, and the width (across the shoulders) is the same as that of the prothorax. The sides of the elytra are a little incurved at the shoulder angles, subparallel from there to beyond the middle, then gradually rounded and converging to the apices, which are slightly divaricate; the sides of the narrowed apical portions are serrate with unequal denticles. The left apical portion has twelve denticles and of these the second, the third and the eleventh are very small; the right apical portion has fifteen denticles, four of them forming two conspicuous double denticles. The innermost apical denticle is small. The outer and the sutural borders are marginate, the outer marginations becoming obsolete near the apex. The elytra are metallic violet about the suture and their outer portions are metallic green.

The short *legs*, which are cyaneous and tinged with metallic green, are lightly punctate with fine golden hairs. The front and middle coxae are globular, the hind coxae transverse. The front tibiae are slightly arcuate and gradually thickened towards the apex, which bears two apical spurs, and the outer margins are crenulate. The front tarsi are wanting, except the first segment which is bilobed. The middle tibiae are straight, with two apical spurs. The hind tibiae are a little curved, with two apical spurs. The first or basal segment of the middle and the hind tarsi is longer than the second and third segments, which are about equal in length; the fourth segment is very short but with long narrow lobes and a large transverse membranous pad; the fifth is long and narrow, gradually broadening to the apex and not lobate, and it bears two simple claws.

The *abdomen* is irregularly and thinly punctulate; the punctures, which are larger and closer together on the lateral parts of the sterna, bear short and fine golden recumbent hairs. The first abdominal sternum is very long; it is connate with the second and the clearly impressed junction line forms a complete border. The second and third sterna are almost equal in length, the fourth is a little shorter than the third, and the fifth is arcuate and rugulose about the rounded apex.

Length 24 mm.; breadth (across the elytra) 8 mm.
Hab. South America (Fab. and Oliv.).
See Plate 35.

The **smaller co-type**, compared with the larger one, shows the following differences:

The head and pronotum and the sutural area of the elytra are metallic bronze green, the outer portions of the elytra are light purple.

The front part of the pronotum has a shallower excavation. The sides of the pronotum (viewed from above) are more markedly angular, the angularity is nearer the middle and the portion of the side anterior to it is incurved, whereas in the larger co-type it is outcurved, thus giving a roughly rounded effect to the whole side.

Length 21 mm.: breadth (across the elytra) 7 mm.

68. *Chrysaspis aurata* (Fab.)

Coleopterorum Catalogus, pars 84 (J. Obenberger, 1926), Buprestidae, I, p. 104. *Monographie des Buprestides* (C. Kerremans), III, 1908–9, p. 109.

Guinea, Belgian Congo, etc.

SYN. *Buprestis aurata* Fab., *Mant. Ins.* I, p. 178, No. 33 (1787); *Ent. Syst.* I, 2, p. 197, No. 49 (1792); *Syst. Eleuth.* II, p. 196, No. 58 (1801); Oliv. *Ent.* II, 32, p. 33, pl. 9, fig. 93 (1790).

Chrysochroa aurata (Fab.), *Catalogus Coleopterorum* (Gemminger and Harold, 1869), v, Buprestidae, etc., p. 1353.
Chrysaspis chrysipennis (Hope), *Catalogus Buprestidarum* (Ed. Saunders, 1871), p. 10.

The specimen under label

'*Bup. aurata*
Fabr. MSS'

in Cabinet B, drawer 8, is the type of this species; it agrees with the descriptions given by Fabricius and with Olivier's description and figure, and it closely resembles the modern examples of *aurata* Fab. in the British Museum Collection.

In his descriptions Fabricius gives America as the habitat, and Olivier states South America, which evidently are errors in locality.

This type was examined by the late Mr Charles O. Waterhouse.[1] He pointed out that it is closely allied to the species known as *elongata* Oliv., that the general form and sculpture of the two species is very similar, but that *aurata* is a slightly broader insect and differently coloured. Other points noted by Mr Waterhouse were: Thorax with scarcely a trace of the coppery colour present in *elongata*; puncturation of the thorax similar in both species. Elytra with a light coppery tint on the outer portion of the apex only; the elytra of *elongata* largely suffused with copper colour, which does not, however, extend to the margins; the puncturation of the elytra distinctly stronger than in *elongata*; the abdomen golden with a light coppery shade as in *elongata*.

Comparing the type with a series of specimens of *elongata* from Acropong, West Africa, in the Bishop Collection, the serrate apices of the elytra of *aurata* are gently rounded in, whereas those of *elongata* are sharply rounded in, consequently the ends of the elytra have a distinctly truncate appearance; and the median epipleural sinuosity is much

[1] 'Observations on Coleoptera of the Family Buprestidae, with descriptions of new species' by Charles O. Waterhouse, F.E.S. (*Annals and Magazine of Natural History*, Vol. XIV, Seventh Series, 1904, p. 345).

less strongly marked in *elongata*. The faint coppery shade on the abdomen is less distinct in *aurata*.

Description of Type, *Buprestis aurata* Fab. The specimen is a *male*. Form elongate and narrow, broadest across the shoulders of the elytra and from there towards the apex somewhat parallel-sided; the disc of the prothorax and the sutural area of the base of the elytra flattened, otherwise moderately convex. Shining; the head and the thorax obscure golden green, the hollowed sides (pleura) clear golden; the elytra bright golden green, the yellow tint dominant, and with the outer parts of the apices very slightly coppery; the epipleura (under the shoulders) violet coloured; the prosternum and the front and middle pairs of legs bright metallic green with a violet tinge in certain lights; the femora of the hindmost legs and the abdomen bright golden with a faint coppery shade. The undersurface of the thorax and abdomen, and also the legs, lightly pubescent with fine recumbent golden hairs.

The *head* is much narrower than the thorax and is strongly hollowed between the eyes, the frons being deeply and (towards the clypeus) broadly furrowed; the clypeus is impressed and widely emarginate. On the short vertex there is a median longitudinal sutural line which is continued deeply on the narrower part of the frons; the vertex is punctate, the frons is partly punctate, partly rugose-punctate, the clypeus is rugose-punctate. The large and elliptical eyes are dull testaceous. The antennae (imperfect) are deep golden green with a violet hue, except the serrate parts of the segments which are dull black.

The *prothorax* is transverse, greatly narrowed in front and there constricted so that the front of the thorax has the form of a raised and convex ring around the very short insunk head. The side margins of the thorax are almost parallel and are rounded in anteriorly towards the annular constriction; the base of the prothorax, which is much wider than the front, is excised and biarcuate; the posterior angles are right angles.

Viewed from above, the side margins of the *pronotum*, which have the form of lateral carinae, appear almost parallel except towards the front where they are rounded in at the proepisternal notch, and there the strong and forwardly curved carina becomes declivous and obsolete. The pronotum is moderately convex, but the disc is flattened. The surface of the pronotum, smooth on the disc and uneven around it, is very finely punctured (punctulate) with minute punctures; and there is also an irregular puncturation of larger punctures, these being stronger and closer together around the disc, particularly towards the front, and lighter and more scattered over the disc. On the disc there is a faint median longitudinal line with two slight but distinct furrow-like impressions, and at each side of the disc posteriorly there is a short and oblique well-marked depression or fovea.

The sides or pleura (*prothoracic episterna*) within the lateral carinae are hollowed and punctulate with a light covering of fine recumbent yellowish hairs.

The *prosternum* is finely and irregularly punctate except along the middle; the punctures are not close together except at the distal end. The *mesosternum* and *metasternum* and the corresponding epimera and episterna are irregularly and very finely puncturate, the punctures being smaller than those on the prosternum. The anterior border and the median longitudinal and median transverse lines of the metasternum are violet coloured.

The *scutellum* is invisible.

The *elytra* are long and subparallel, gradually narrowed and rounded in at the apex, a little broader than the thorax, broadest across the shoulders, and flattened over the sutural area of the base. The outer borders of the elytra are slightly sinuate with a distinct angular infra-humeral projection; beyond this projection the outer border is deflexed as far as the hind coxae, there forming a strongly marked bend (median epipleural sinuation) as seen from beneath. The outer borders are marginate, except the apical portions which are

Buprestis aurata Fab. ♂ × 3

serrulate with the sutural tips acuminate; the inner (sutural) borders are slightly marginate posteriorly.

The elytra are lightly sculptured on their outer portions, where the costae are slight but distinct and where, between the prominent humeral callus and the outer border, the surface is deeply furrowed and rugose-punctate. The surface of the elytra is, like that of the thorax, very finely punctured with minute punctures; and there are numerous larger punctures in regular punctate-striate formation, in the form of double rows, and also irregularly in the intervals or interstices. On each elytron there are four double rows of punctures representing the borders of four costae; the punctures of the outermost double row are faintly marked towards the apex and form the border of a slightly convex but well-defined costa, which extends from the prominent shoulder callus to the apex and parallel with the outer border; the next two double rows of strongly marked punctures become obsolete towards the apex at a point where the two intermediate and less distinct costae become confluent and are continued as one. The inner intermediate costa is scarcely convex, except posteriorly where it unites with the outer one. The innermost double row of punctures, less strongly marked, extends to near the apical angle, where the costa which it represents becomes evident as such and is united with the posterior continuation of the two intermediate costae. Between the innermost double row and the suture there is a single subsutural row of lightly marked punctures and also a short (scutellary) double row confluent at the suture. The three costal intervals or interstices are irregularly punctured except about the apex. Towards the outer sides and over the shoulders, where the punctures tend to become confluent, there is a slight rugulose sculpturing.

The *legs* are bright deep metallic green; the femora, tibiae and tarsi are strongly punctured, the punctures bearing hairs, but the hindmost femora have a bright golden hue and the puncturation is finer and much closer. The tarsi have the segments lobed and the tarsal claws are simple; the hindmost tarsi are more than half the length of the tibiae.

The *abdomen* is lightly punctured (punctulate), the punctures are smaller and closer together on the sides of the sterna and each puncture bears a fine recumbent light golden hair. The fifth abdominal sternum is widely but not deeply emarginate.

Length 33 mm.; breadth (across the elytra) 11 mm.
Hab. America (Fab.), South America (Oliv.).
See Plate 36.

69. *Halecia blanda* (Fab.)

Coleopterorum Catalogus, pars 84 (J. Obenberger, 1926), Buprestidae, I, p. 148. *Monographie des Buprestides* (C. Kerremans), III, 1908–9, p. 358.

Brazil, Argentina.

SYN. *Buprestis blanda* Fab., *Sp. Ins.* I, p. 276, No. 23 (1781); *Mant. Ins.* I, p. 178, No. 32 (1787); *Ent. Syst.* I, 2, p. 197, No. 48 (1792); *Syst. Eleuth.* II, p. 196, No. 57 (1801); Oliv. *Ent.* II, 32, p. 93, pl. 9, fig. 94 (1790).

The specimen under label

'*Bup. blanda*
Fabr. pag. 276, No. 23'

in Cabinet B, drawer 8, is evidently the type; it corresponds to the descriptions given by Fabricius and Olivier, and Olivier's figure closely resembles it.

Obenberger places this insect under the Genus *Halecia*; K. G. Blair considers it to be *Pelecopselaphus elongatus* Thomson.

Description of Type, *Buprestis blanda* Fab. Form roughly elongate-oval and sharply narrowed posteriorly; but the outline of the posterior half of the thorax and the anterior half of the elytra is continuous and almost straight. Broadest across the shoulders of the elytra and there slightly sinuous, sharply narrowed behind the middle of the elytra to the obliquely truncate apices. The disc of the thorax convex, but with marked depressions, the base of the elytra flattened around the suture. The head coppery and violet with bronze green tinges, and partly rugose-punctate; the thorax violet coppery tinged with bronze green and irregularly punctate; the elytra violet coppery with tinges of bronze green, costate and punctate-striate; the under surface of the body bright metallic green, suffused with violet on the middle parts of the sterna; the legs tinged with violet.

The *head* is narrower than the front of the thorax; the vertex, somewhat expanded behind the eyes, is convex, punctate, coppery tinged, and it is marked by a median longitudinal impressed line, which is interrupted at a small bridge-like elevation (rugosity) between the vertex and the frons; the frons is hollowed, rugose-punctate, tinged with violet, and traversed by the continuation of the median line, which becomes obsolete before the clypeus at a line drawn immediately above the insertions of the antennae; the clypeus is depressed, punctate, tinged with violet and widely emarginate about the base of the labrum, which is narrow, semicircular, punctate and covered with a thick fringe of long yellowish hairs.

The large and elliptical eyes are violet bronze. The antennae are inserted in the triangular areas immediately in front of the eyes and above the front angles of the clypeus; the first and longest antennal segment is club-shaped, the second and shortest is rounded and knob-like, the third segment is larger than the eight successive serrate segments, but its shape is similar; the first three antennal segments are tinged with violet, the serrate segments are dull black and they have small kidney-shaped pits or foveae containing the antennal pores.

The *prothorax* is transverse, its breadth is almost one-third greater than its length; it is broadest behind the middle and is obliquely narrowed towards the front where it is margined by a thickened ring. This ring is interrupted behind the eye, at the temple and in front of the prothoracic episterna, and it is flattened above (at the raised median portion of the front of the pronotum) and there flush with the surface of the disc.

The *pronotum* is excavate and sinuate in front, its base is sinuate with a small scutellar lobe, its sides (viewed from above) are gradually rounded from the base towards the front and there is a very slight sinuation just above the right-angled posterior angles; the sides are marginate (lateral carinae) and the margins or carinae are deflected towards the front and are there continuous with the interrupted frontal thickened ring, forming marked anterior angles.

The pronotum is convex and its surface is irregularly punctate; the punctures diminish both in number and size about the middle of the disc, particularly about the base. At the base, in front of the small scutellar lobe, there is a median longitudinal depression in the middle of which there is a short and isolated longitudinal line; under magnification this line has an oval beginning and forms a deeply impressed sulcus, anteriorly it is lightly impressed and gradually fades out towards the centre of the pronotal disc. At each side of the disc there is an oblique and roughly oval impression flanked by ridge-like elevations. At the base of the pronotum, and near the posterior angle each side, there is a semi-circular depression, in conjunction with a similar depression on the base of the elytron in front of the shoulder callus.

The sides or pleura (*prothoracic episterna*) are broadly triangular, hollowed, and irregularly punctate, the punctures bearing short and recumbent light yellowish hairs.

The *prosternum* is somewhat flattened and is irregularly punctate; the punctures, which bear light yellowish hairs, are closer together on the sides and become fewer and smaller on the prosternal process towards its rounded apex.

The *mesosternum* and the *metasternum* are punctate; the punctures, which bear light yellowish hairs, are fine and scattered on the middle portions of the sterna, they are coarser and closer together on the side portions and they become more or less confluent, with rugulose effect, on the epimera and the episterna.

The *scutellum* is small, almost square with rounded corners, smooth and with a very slight concavity.

Buprestis blanda Fab. × 6

The *elytra* are two and a half times as long as broad and are (across the shoulders) slightly wider than the prothorax; the basal portions are somewhat flattened about the suture; the sides are sinuate about the shoulders, thence subparallel to beyond the middle and there sharply bent in and convergent to the obliquely truncate apices; the sides of the narrowed apical portions are rather finely serrate with seven subequal denticles (the first two very small and blunt, the succeeding five acute), and the apices are acutely tridentate with the innermost denticle conspicuously large. The outer and the sutural borders of the elytra are marginate. On each elytron there are five costae, roughly parallel and markedly convex except on the basal area. The subsutural costa, which shows two interruptions at the base, extends to the tip of the apex and ends at the large innermost denticle; the second costa extends to the apex and ends near the middle denticle; the third costa becomes confluent with the second near its termination; the forth costa, arising from the shoulder callus, is short, it becomes obsolete beyond the middle of the elytron and within the sharp loop formed by the fifth costa joining the terminal portion of the third costa. The smooth surface of the costae is lightly marked with some fine punctures and with slight cross-striation here and there. The first costal interstice is in greater part impunctate; the other interstices are irregularly punctate-striate, the punctures becoming fewer towards the apex.

The *legs*, which are dark metallic green tinged with violet, are lightly and not closely punctate with fine yellowish hairs. The front tibiae are arcuate and gradually thickened towards the apex; the inner margin is slightly crenulate and along the inner side there are four small tubercles at irregular distances apart. The first three segments of the front tarsi are lobed, the fourth segment is very small and deeply cleft, the fifth is lobate and the claws are simple. The middle and hind tibiae are almost straight, with two apical spurs; the first or basal segment of the hind tarsus is longer than the second and third segments together, but it is less than half the length of the entire tarsus.

The *abdomen* is irregularly punctate, the punctures bearing short, fine and yellowish recumbent hairs; the lateral parts of the sterna are closely punctate, but on the middle parts the punctures are fewer and scattered. The first and second abdominal sterna are fused in the middle, the fifth is arcuately emarginate, and it is closely punctate except on a portion of the middle strip.

Length 20 mm.; breadth (across the elytra) 6 mm.
Hab. South America (Fab. and Oliv.).
See Plate 37.

70. *Chrysesthes tripunctata* (Fab.)

Coleopterorum Catalogus, pars 84 (J. Obenberger, 1926), Buprestidae, I, p. 154. *Monographie des Buprestides* (C. Kerremans), III, 1908–9, p. 267.

Honduras, Cayenne, Brazil, Argentina, etc.

SYN. *Buprestis 3-punctata* Fab., *Mant. Ins.* I, p. 179, No. 34 (1787); *Ent. Syst.* I, 2, p. 198, No. 51 (1792); *Syst. Eleuth.* II, p. 197, No. 62 (1801); Oliv. *Ent.* II, 32, p. 42, pl. 2, fig. 10 (1790).
Chrysesthes tripunctata var. *ambigua* Gory.
C. steinheili Thomson.
C. impunctata Théry.

This species is represented by one specimen in Cabinet B, drawer 8, under label

'*Bup. tripunctata*
Fabr. MSS'

which is the type; it answers the descriptions given by Fabricius and Olivier and corresponds to Olivier's figure, and it closely resembles the modern examples of *tripunctata* in the British Museum.

Description of Type, *Buprestis tripunctata* Fab. Form elongate-oval, broadest behind the shoulders of the elytra and from there sinuous and tapering to the elytral apex; the prothorax moderately convex, the elytra flattened. Metallic coppery green above, with three small insunk golden spots on each elytron: lightly pubescent and bright golden green beneath, except the terminal part of the last abdominal sternum which is bronze green tinged with violet. The head and the thorax purplish coppery green.

The *head* is slightly broader than the front of the thorax, it is slightly excavated between the eyes and it is marked by a median longitudinal furrow which is linear on the vertex and irregularly foveate on the frons and which becomes obsolete before the clypeus; the clypeus is impressed and widely emarginate. The vertex is punctate; the frons is punctate, some of the punctures are confluent and golden and about the eyes the frons is golden; the clypeus is punctate and rugulose and golden about the front; the labrum is narrow, semicircular, deeply emarginate in front, and punctate with whitish hairs. The large, elliptical and finely faceted eyes are purplish bronze green suffused with brighter green. The antennae are inserted in the golden triangular areas between the lower parts of the eyes (in front) and the front angles of the clypeus and are lodged in shallow infra-orbito-genal grooves which are golden.

The *prothorax* is transverse, the breadth being about one-third greater than the length; it is broadest at the base and narrowed towards the front, where it is margined by a slightly thickened ring which is partly obsolete above on the elevated median portion of the front of the pronotum.

The *pronotum* is straight in front, its base is sinuate and is elevated in greater part, particularly the scutellar lobe; the sides of the pronotum are sinuate and have the form of lateral carinae, which, however, become obsolete towards the front; the posterior angles are right angles. The pronotum is moderately convex, the surface is irregularly punctate; at the base, in front of the scutellar lobe, there is a median depression marked by a short, longitudinal, deeply impressed line.

The sides or pleura (*prothoracic episterna*) are broadly triangular, a little hollowed and irregularly punctate, excepting a strip beneath each carina which is impunctate and the anterior half of which is coppery.

The *prosternum* is flattened and is irregularly and finely punctate, anteriorly the punctures are linear and confluent with a slight rugulose effect and posteriorly they diminish both in size and number towards the rounded apex of the prosternal process.

The *mesosternum* and the *metasternum* are finely punctate with scattered punctures bearing short and fine golden hairs; the punctures are closer together on the epimera and episterna and on the lateral portions of the sterna.

The *scutellum* is bronze green, small, irregular in form, rounded and slightly convex; and it is transverse, its breadth being twice its length.

The *elytra* are rather flattened, sinuate at the sides and fully twice as long as wide, wider than the prothorax; they are broadest across the rounded shoulders, incurved near the middle, almost parallel in the middle third and convergent in the apical third, gradually curving in towards the acuminate apex (sutural spine); and the sides of the apical area are serrate with eleven acute subequal denticles. The outer and sutural borders are marginate, excepting the basal third of the latter. Between the shoulder callus and the suture there is a wide basal depression irregular and partly rugulose. The costae, largely parallel and ten in number on each elytron, are very slightly convex anteriorly and show more distinct elevation on the middle and apical areas; the first costa is rather indistinct and short, becoming confluent with the second (subsutural) costa which is obsolete towards the apex; the third extends to the apex, where it apparently unites with the tenth; the fourth and the sixth costae unite near the apex in a very distinct sharp loop, and the fifth joins the sixth near its junction with the fourth; just beyond the posterior fovea the seventh, eighth and ninth costae become

Buprestis tripunctata Fab. × 4

terminally confluent. The costae are mainly smooth, there being few punctures on their surfaces, and the costal interstices are punctate-striate with irregular double rows of golden punctures in the three inner and the two outer interstices and irregular single rows in the three middle interstices; the irregularity of the punctures is very marked in the outer interstices, particularly about the middle, where they are scattered on the costae. On each elytron there are three small golden pits or foveae arranged in a median longitudinal line at almost equal distances apart; the middle one (double on the left elytron), which is the smallest of the three, is centrally situated on the seventh costa and nearly half the length of the elytron from its base; the anterior one, a little larger and on the fifth costa, is equidistant between the middle fovea and the base of the elytron; the posterior fovea is the largest and is double, it consists of two conjoined aggregates of fine golden punctures situated on the sixth and seventh costae and nearly equidistant between the middle fovea and the apex of the elytron.

The *legs*, which are bright golden green, with the upper surfaces of the tibiae and tarsi darker bronze green, are lightly and not closely punctate, the punctures bearing fine yellowish hairs. The front tarsi have the first three segments lobed, the fourth segment is deeply cleft, the fifth is lobate and the claws are simple.

The *abdomen* is irregularly and finely punctate; the punctures, which bear short and fine recumbent yellowish hairs, are closer together on the lateral parts of the sterna and are a little larger and scattered on the middle parts. The fifth abdominal sternum is arcuately emarginate, the ends of the arch forming conspicuous spine-like processes; its surface is closely punctured, and the terminal portion is dark bronze green tinged with violet.

Length 22 mm.; breadth (behind the shoulders of the elytra) 8 mm.
Hab. South America (Fab.), Cayenne, Surinam (Oliv.).
See Plate 38.

71. *Psiloptera* (sub-genus *Damarsila*) *punctatissima* (Fab.)

Coleopterorum Catalogus, pars 84 (J. Obenberger, 1926), Buprestidae, I, p. 179. *Monographie des Buprestides* (C. Kerremans), V, 1910, p. 232.

Congo, Senegal, Guinea, Liberia.

SYN. *Buprestis punctatissima* Fab., *Syst. Ent.* p. 217, No. 5 (1775); *Sp. Ins.* I, p. 274, No. 6 (1781); *Mant. Ins.* I, p. 176, No. 6 (1787); *Ent. Syst.* I, 2, p. 188, No. 10 (1792); *Syst. Eleuth.* II, p. 188, No. 12 (1801); Oliv. *Ent.* II, 32, p. 14, pl. 7, fig. 76 (1790).

This species is represented by three specimens in Cabinet B, drawer 8, under label

'*Bup. punctatissima*
Fabr. pag. 274, No. 6'

and they correspond with the descriptions given by Fabricius and Olivier and with Olivier's figure of *punctatissima*. One of the specimens has been compared with modern examples of the species in the British Museum Collection and it closely agrees.

Fabricius first described this species from a specimen belonging to Drury; a later description in *Ent. Syst.*, repeated in *Syst. Eleuth.*, is from an example in the Banks Collection. Probably the Drury insect was acquired by Hunter. Olivier's description is from a Hunterian example.

Description of Type, *Buprestis punctatissima* Fab. Form oblong-obovate, the outline almost continuous; the sides in greater part subparallel and a little sinuous; slightly broader across the shoulders of the elytra than elsewhere and rather sharply rounded off and narrowed towards the bidentate elytral apices; convex, slightly flattened over the disc of the prothorax, the dorsal surface (minutely areolate) sculptured with rugose elevations and small foveae or pits (on the head) and also punctures, closely punctate-striate on the elytra; and with a fringe of short hair on the front border of the pronotum, thinly continued and scattered along the sides of the pronotum and the elytra to the apices. Two short impressed lines, roughly crescentic and obliquely placed near together on the scutellar part of the prothorax. Brilliant golden green above with numerous scattered irregular and violet-blue-coloured spots, mostly oblong on the elytra and these mostly in regular longitudinal rows. Mainly bright coppery beneath, and with a few violet spots; the surface of the underparts and the legs irregularly punctate, the punctures bearing recumbent yellowish hairs.

The *head* is narrower than the front of the thorax; it is bright golden green with some irregular violet-blue markings. The sculpture of the head is variolose; it is closely pitted with small round and oval foveae, some of which are conjoined or confluent, and it is rugose, the interspaces being elevated. Under magnification the entire surface (foveae and interspaces) is minutely areolate. The vertex is convex and broad, a little expanded behind the eyes, and, extending from the base to the frons, there is a finely impressed violet-coloured median line. The frons is slightly hollowed before the clypeus; on the frons at each side, and extending to the emargination of the clypeus, there is an obliquely placed and crescentic ridge which marks off the hollowed antennal area in front of each eye. The clypeus is thickened and arcuately emarginate in front in the middle, around the base of the labrum; and the labrum is narrow, semicircular, punctate and covered with a thick fringe of long yellowish hairs. The bases of the mandibles and the mentum are closely punctate, the punctures bearing hairs. The elliptical eyes, fulvous, suffused with dark brown and with a thin purple border, are narrowed, not particularly prominent, not close together, and they are obliquely placed. The antennae, inserted in the hollow triangular fronto-clypeal area in front of the eyes, are bright bronze green; the first and longest antennal segment is club-shaped, the second and third segments are knob-like, nearly equal in size, and smaller than the succeeding serrate segments, which have double and conjoined rounded pits or foveae containing the antennal pores.

The *prothorax* is transverse, its breadth is nearly one-third greater than its length, it is broadest across the middle. The *pronotum* is slightly but widely excavate in front; the base, which is one-third wider than the front, is sinuate corresponding to the arched bases of the elytra and the scutellar lobe is very small; the sides are evenly rounded, the posterior angles are a little projecting, and the lateral carinae are represented by blunt ridges, which are slightly deflected and become obsolete towards the front. The pronotum is convex, somewhat flattened over the disc and slightly hollowed about the middle of the base; its surface, minutely areolate and with some irregular violet-blue-coloured markings, is irregularly and closely punctate with rugose interspaces. The scutellar part of the base is clearly marked off as a lozenge-shaped impunctate area by two short deeply impressed and roughly crescentic lines, near together and obliquely placed.

The *prothoracic episterna* are broadly triangular, partly convex and closely variolose with large round and shallow punctures bearing recumbent yellowish hairs.

The *prosternum* is irregularly punctate with oval punctures bearing recumbent yellowish hairs. The coppery coloured prosternal process is flattened, it has two longitudinal sulci, each containing a row of oval punctures mostly confluent and with hairs; and its tip, which reaches to the metasternum, is obtusely triangular behind the anterior coxal cavities.

The *mesothoracic epimera* and *episterna* are hollowed and variolose with round shallow hair-bearing punctures. The *mesosternum* is punctate, except on the oblique ridge-like elevations which bound the central hollow and which are smooth.

The narrow and triangular *metathoracic episterna* are closely punctate. The *metasternum* is variolose on the sides with large

Buprestis punctatissima Fab. × 5

round and shallow punctures which have yellowish recumbent hairs; it is punctate towards and on the middle where the punctures are smaller, oval-shaped and more scattered. The coppery-coloured ante-coxal portion of the metasternum is clearly marked off by a thin violet-coloured arcuate line, the ends of which do not quite reach the base, and it is divided by a median longitudinal suture; this suture merges at the violet boundary line with a rather broad and deep sulcus, containing oval punctures, which becomes narrowed into a line extending to the front of the metasternum.

The *scutellum* is small and round, with a few fine punctures.

The *elytra* are convex, slightly flattened about the suture at the base, almost two and a quarter times as long as broad and, across the shoulders, slightly wider than the prothorax. The bases of the elytra are arcuate and the shoulder angles are sharply rounded. The sides of the elytra are somewhat sinuate along the shoulder region, where the margin is represented by a rugose thickening above the narrow hollowed epipleura, which overlap the metathoracic episterna; along the middle area the sides are subparallel and from there rather sharply rounded and converging to the narrow truncate, emarginate and bidentate apices, which are distinctly divaricate. The margin of the outer borders of the elytra is represented by an irregular ridge with punctures bearing hairs; the suture is not marginate.

The surface of the elytra is sculptured like that of the prothorax; but the close puncturation is different, it is markedly regular, the punctures forming a close series of longitudinal and mainly parallel striae of two kinds, one composed of a single row of large coarse punctures close together and the other consisting of a row of small ones farther apart. These striae are arranged alternately and the punctures vary in shape from round to oval. On the outer sides of the elytra the rows of punctures become irregular and confluent with rugose effect. The violet-blue-coloured spots on the elytra also show regularity in arrangement and tend also to regularity of shape, for they are mostly in definite rows and are mostly oblong, roughly rectangular.

The *legs* are bright metallic golden green, punctuate and hairy. The front tibiae are a little arcuate, gradually thickened towards the apex, which has an outer and triangular apical process. The first four segments of the front tarsi are lobed; the fourth segment is smaller than the preceding three segments and is deeply cleft, the fifth segment is lobate and the claws are simple. The middle and the hind tibiae are straight, with two apical spurs. The first three segments of the middle and hind tarsi are about equal in length.

The *abdomen* is coppery about the middle with a few violet spots of irregular shape on the five sterna; its surface is rugose and is irregularly and coarsely punctate with oval-shaped punctures, which tend to coalesce in little groups and which bear short yellowish recumbent hairs. The first abdominal sternum is longer than the second and its middle portion is dagger-shaped, there being two roughly parallel blunt ridges which converge between the posterior coxae to a sharp point and which enclose a hollow containing oval punctures, smaller and closer together within the converging part. The junction of the connate first and second sterna is a clearly marked entire border. The third and fourth sterna are about equal in length and shorter than the second. The fifth sternum is truncate and very slightly emarginate; it is also fimbriate, the hairs of a row of coalesced punctures forming a fringe around the apical portion.

Length 20 mm.; breadth (across the elytra) 6½ mm.
Hab. Sierra Leone (Fab. and Oliv.).
See Plate 39.

72. *Aglaostola tereticollis* (Pallas)

Pallas, *Icones Insectorum*, 1782, pp. 75–76, t. D, f. 18. *Coleopterorum Catalogus*, pars 111 (J. Obenberger, 1930), Buprestidae, II, p. 362. Wytsman, *Genera Insectorum*, Buprestidae (C. Kerremans), Fasc. XII, 1903, pp. 155–156. *Catalogus Buprestidarum* (Ed. Saunders, 1871), p. 47.

Jamaica.

SYN. *Buprestis corrusca* Fab., *Mant. Ins.* I, p. 176, No. 8 *corusca* (1787); *Ent. Syst.* I, 2, p. 188, No. 13 (1792); *Syst. Eleuth.* II, p. 189, No. 17 (1801); Oliv. *Ent.* II, 32, p. 17 *corusca*, pl. 9, fig. 99 (1790).

The type of this species is, as stated by Fabricius, in the Banks Collection, British Museum. The single specimen in Cabinet B, drawer 8, under label

'*Bup. corrusca*
Fabr. MSS'

is a *male*; it closely agrees with the type and with modern examples in the British Museum.

Professor Graham Kerr indicated[1] this Hunterian specimen as possibly one compared by Fabricius with the type and regarded by him as conspecific with it, therefore a metatype.

Description of Metatype, *Buprestis corrusca* Fab. Form narrow elongate-obovate, continuous in outline, broadest across the shoulders of the elytra, sinuate at the sides of the elytra and gradually rounded in to the narrow, truncate and bidentate apices; the disc of the thorax not very convex, the sutural area of the elytra rather flattened. Smooth and shining; the head and thorax finely and irregularly punctate, the elytra faintly punctate, the punctules in lines (punctulate-striate) towards the suture and very faint and scattered about the sides. Coloration of the head and thorax brilliant golden green, the eyes brown with red, the antennae dark metallic green and violet blue; the elytra bright bronze green with touches of violet blue at the shoulders and about the middle of the suture, and with suffusion of coppery red over the apices. Under surface of the body and the femora golden green, the upper parts of the femora suffused with violet blue.

The short insunk *head*, slightly narrower than the front of the thorax, is convex over the vertex and flattened in front;

[1] 'Remarks upon the Zoological Collection of the University of Glasgow' by Professor J. Graham Kerr (*Glasgow Naturalist*, Vol. II, No. 4, September 1910, p. 106).

it is bright golden green and its surface is finely punctate with scattered punctures. The vertex is convex, it is marked by a rather faint median longitudinal line which extends from the base to the front and which traverses a small depression between the eyes. The frons, slightly convex, widens out towards the antennal insertions. The clypeus, which is hollowed and coarsely punctate, is very short and thickened, emarginate in the middle in front (over the labrum) and with two crescentic ridges obliquely placed and overarching the hollowed antennal areas at the lower or front ends of the eyes. The labrum is semicircular, punctate and fringed with yellowish hair; the mentum is transverse and punctate. The oval eyes are fairly prominent, placed vertically oblique and not close together; the finely faceted surface is brown with an inner ring of red. The antennae are inserted under the oblique clypeal ridges, at a short distance from and on a level with the lower ends of the eyes. The first two antennal segments are bright golden green; the first or basal segment is long and clavate, the second is short and knob-like, the succeeding segments are serrulate and dark bronze green tinged with violet blue.

The *prothorax* is transverse, its basal breadth is one-third greater than its length; narrowed towards the front, it is broadest at the base. The front is slightly thickened, but the ring-like constriction is well marked at each side only. The *pronotum* is nearly straight in front; its base is somewhat sinuate with a slight but wide median lobe. The sides are slightly curved (viewed from above) and considerably inflected towards the front; and so the side margins, which are raised (carinate) and also sinuate, are directed obliquely forwards from the sharp posterior angles. The pronotum is marginate on the base, except about the middle, and is slightly and unevenly convex, there being five distinct impressions upon the disc, located as follows: a shallow median-basal impression and, between it and the angle of each side, a deep puncture-like depression; and directly in front of each deep depression there is a light impression. The pronotal surface is irregularly punctulate and brilliant golden green.

The *proepisterna* and *proepimera* are broadly triangular and convex, the narrower and hollowed *mesepisterna* and *mesepimera* are roughly triangular with one side curved; and these parts, including the oblong *metepisterna*, are thinly punctate, the punctures bearing short and fine golden hairs. The convex *prosternum* is coarsely punctate, and the short triangular prosternal process is lodged between the elongate lateral parts of the *mesosternum*. The *metasternum* is thinly punctulate; the ante-coxal parts are marked off anteriorly by a roughly arcuate line, the ends of which do not quite reach the base of the metasternum. A median longitudinal suture extends from the base to the front of the metasternum. The punctures of the thoracic sterna bear short and fine golden hairs.

The *scutellum* is small and convex, with its rounded apex inflected.

The *elytra* are narrow and long, the length is fully two and a half times the breadth, which is a little more than that of the prothorax; the shoulder angles are obtuse and the shoulder callus is prominent; the sides are subparallel from the shoulders to the apical third and from there gradually rounded in and converging to the narrowed apices, which are obliquely and arcuately truncate and bidentate, and the outer denticle is larger than the inner or sutural point. The basal third of the elytra is considerably inflected at each side and also hollowed above the narrow epipleural part; the remaining two-thirds of each side is less inflected and is marginate, and the suture is marginate; the outer edges are sinuate (viewed from the side). The elytra are bright bronze green suffused with violet blue about the shoulders and the middle of the suture, and with coppery red on the apical parts. The elytral surface is finely punctate and also strigose on the lateral parts; it is punctulate-striate towards the suture, where the punctules form four or five distinct parallel lines, and about the sides the punctules are rather faint and scattered.

The front and middle coxae are globular; the hind coxae are transverse, with their posterior margins oblique and strongly sinuate and bordered by a grooved line. The short *legs* are thinly punctulate, the punctules bearing fine yellowish hairs; all the femora are stout, the front tibiae are arcuate, the middle and hind tibiae are straight. The distal spines of the middle and hind tibiae are subequal and larger than those of the front tibiae. In size and shape the segments of the front tarsi are similar, except the last, which is longer and clavate. The first segment of the middle tarsi is nearly as long as the fifth, which is narrow and clavate; the second, third and fourth segments are shorter and broader progressively and the fourth is strongly bilobed. The segments of the hind tarsi are similar to those of the middle tarsi; but the first segment is the longest, it is clavate and almost equal in length to the second and third segments together. All the tarsal segments, except the fifth, bear membranous pads, and the claws are simple.

The first sternum of the *abdomen* is very long, twice the length of the second, and its front portion forms a prominent and pointed median projection between the hind coxae. The second sternum is longer than the third, the fourth is the shortest, the fifth is as long as the second and is notched at the narrowed end. The suture of the connate first and second segments is distinct. The surface of the abdominal sterna is thinly and faintly punctulate.

Length 14 mm.; breadth (across the shoulders of the elytra) 4¼ mm. Hab. Jamaica (Fab. and Oliv.).

73. *Buprestis fasciata* Fab.

Coleopterorum Catalogus, pars 111 (J. Obenberger, 1930), Buprestidae, II, p. 392. *Catalogue of the Coleoptera of*

America, North of Mexico (Charles W. Leng, 1920), p. 181. Wytsman, *Genera Insectorum*, Buprestidae (C. Kerremans), Fasc. XII, 1903, p. 142.

Canada, Nova Scotia, U.S.A.

SYN. *Buprestis fasciata* Fab., *Mant. Ins.* I, p. 177, No. 13 (1787); *Ent. Syst.* I, 2, p. 191, No. 23 (1792); *Syst. Eleuth.* II, p. 191, No. 31 (1801); Oliv. *Ent.* II, 32, p. 21, pl. 9, fig. 92 (1790).

In Cabinet B, drawer 8, there are two specimens under label

'*Bup. fasciata*
Fabr. MSS'

Evidently the larger one is the example on which Fabricius founded this species, it agrees exactly with his description. Both specimens answer to the description given by Olivier; but his figure is not a good representation of either. This type has been compared with modern examples of *fasciata* in the British Museum Collection.

Description of Type, *Buprestis fasciata* Fab. Form elongate-oval, broadest across the shoulders of the elytra, with a marked sinuosity on the sides of the elytra, and gradually rounded in towards the truncate apices; the disc of the prothorax moderately convex, the sutural area of the elytra slightly flattened. Shining metallic bronze green, but with a strong suffusion of violet blue on the head and thorax and of violet on the sutural and apical area of the elytra. The surface of the head and thorax irregularly punctate, the elytra costate and punctate-striate, with orange-coloured band-like markings. The undersurface of the body bright metallic bronze green strongly tinged with violet blue, irregularly punctate and lightly pubescent with fine golden hairs; the legs bluish green, punctate and lightly covered with fine hairs.

The *head* is narrower than the front of the thorax and it is convex; the frons is widened out at the insertions of the antennae and is convex towards the clypeus, which is widely emarginate about the base of the labrum. The labrum is bronze green and transversely impressed, it has two basal foveae, some punctures anteriorly, and a distal fringe of fine golden hairs. On the short vertex there is a lightly impressed median longitudinal sutural line which becomes obsolete at the frons. The vertex is punctate, the frons and the clypeus are rugulose punctate. The large and elliptical eyes are dull testaceous. The first three segments of the antennae are bronze green; the eight succeeding segments, which are slightly serrate and blackish green, have inferior terminal foveae in which the antennal pores are concentrated.

The *prothorax* is transverse; it is broadest behind the middle and obliquely narrowed towards the front, where it is margined with a thickened ring. The *pronotum* is slightly sinuate in front and strongly sinuate behind; the side margins, which take the form of lateral carinae, are curved in near the front and there the carinae become obsolete; it is moderately convex, its surface is irregularly punctate, and at the posterior angles, which are right angles, there is a slight fovea. The sides or pleura (*prothoracic episterna*) are convex and irregularly punctate.

The *prosternum* is flat and is irregularly and strongly punctate, and the punctures, which bear short and fine golden hairs, diminish both in size and in number towards the apex of the prosternal process. The *mesosternum* and the *metasternum* are strongly punctate; the punctures are contiguous on the episterna and epimera and on the lateral parts of the sterna and they bear fine recumbent golden hairs; towards the middle the punctures are fewer and scattered. The suture between the metathoracic episternum and the metasternum is strongly carinate.

The *scutellum* is small, longer than broad, with the apex widened out and rounded; it is centrally hollowed and it has two anterior punctures.

The *elytra* are long, almost three times as long as wide, somewhat flattened and with sinuate outer borders; they are broadest across the rounded shoulders, widely incurved about

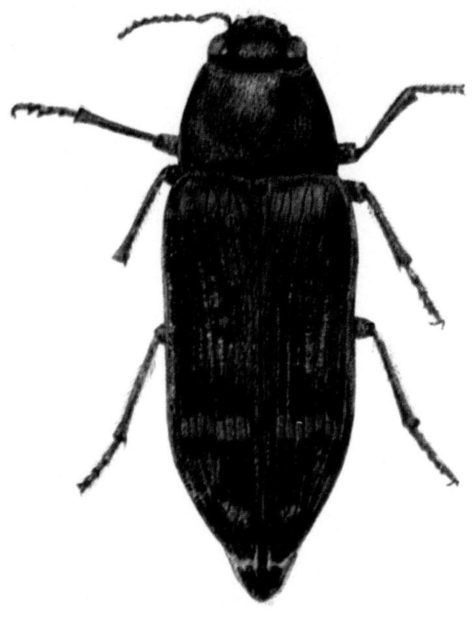

Buprestis fasciata Fab. × 7

the middle, rounded posteriorly and gradually curved in towards the truncate and bidentate apex; the apices are slightly divergent (dehiscent) and the two apical teeth are small and not acute. The outer and sutural borders of the elytra are marginate. The costae, largely parallel and eleven in number on each elytron, show more prominent elevation on the apical area; the first costa extends only a short distance, becoming confluent with the subsutural costa, which is obsolete near the sutural tooth; the next four costae are joined together by branch connection in the apical area; the seventh, eighth and ninth unite in a loop posteriorly and are continued as one in the apical area; the eleventh costa is distinct as such posteriorly, but it becomes obsolete before the apical area. On each costa there is an irregular row of scattered punctures, and between the costae the elytra are punctate-striate.

Near the middle of the left elytron, on the third and fourth costae, there is a very small orange-coloured spot (present on both elytra of the co-type); and on each elytron, behind the middle and in front of the loop junction formed by costae 7 to 9, there is a transverse wavy orange-coloured band or fascia, which does not reach the sutural and outer borders, and within the apical area there is a similar but much smaller orange-coloured fascia.

The *legs*, which are metallic bluish green with deep metallic green tarsi, are lightly and not closely punctate, the punctures bearing fine golden hairs; the tarsi have the first two segments lobate, the third segment is plain, the fourth segment of the middle and hind tarsi is emarginate; the claws are simple.

The *abdomen* is irregularly punctate and the punctures, which are closer together on the sides of the sterna, bear short fine recumbent golden hairs. The fifth abdominal sternum is not emarginate.

Length 15 mm.; breadth (across the elytra) 5 mm.
Hab. North America (Fab. and Oliv.).
See Plate 40.

74. *Chrysobothris dorsata* (Fab.)

Coleopterorum Catalogus, pars 132 (J. Obenberger, 1934), Buprestidae, III, p. 593. *Catalogus Coleopterorum* (Gemminger and Harold, 1869), V, p. 1425. *Catalogus Buprestidarum* (Ed. Saunders, 1871), p. 100, No. 133. Wytsman, *Genera Insectorum*, Buprestidae (C. Kerremans), Fasc. XII, 1903, p. 189.

Africa.

SYN. *Buprestis dorsata* Fab., *Mant. Ins.* I, p. 179, No. 38 (1787); *Ent. Syst.* I, 2, p. 199, No. 56 (1792); *Syst. Eleuth.* II, p. 198, No. 68 (1801).
Buprestis serrata Fab., *Ent. Syst.* I, 2, p. 200, No. 61 (1792); *Syst. Eleuth.* II, p. 199, No. 75 (1801).

The specimen under label

'*Bup. dorsata*
Fabr. MSS'

in Cabinet B, drawer 8, is evidently the type; it answers the original description given by Fabricius in his *Mantissa Insectorum* and amplified in the *Systema Eleutheratorum*, and it has been compared with modern examples of this species in the British Museum Collection.

The localities for *dorsata* Fab. are African; Fabricius gives South America, apparently an error.

Description of Type, *Buprestis dorsata* Fab. Form elongate-oblong, broadest across the base of the elytra, which are rather short and apically narrow, with angulate bases and with the outer sides slightly angular and also extensively serrulate. The thorax gently and evenly convex and closely punctulate, becoming rugulose at the sides; the elytra unevenly convex, each with three basal depressions, with faint indications of costae and closely punctulate. The head, thorax and elytra dark metallic bronze green with touches of violet on the thorax and along the elytral suture; the upper surface of the abdomen (exposed to view) bright golden. The under parts of the body uniformly dark bronze green, punctulate with short and fine recumbent hairs, and also marked with fine channelling on the lateral portions of the abdominal sterna; legs dark bronze green, the anterior femora very stout, marked with fine channelling and each bearing a short but very conspicuous lateral spine.

The short *head* is broad, as wide as the front of the thorax; it is deeply hollowed and overarched in front and it is dark metallic bronze green; its surface is closely punctate with irregular punctures, except the frontal area which is rugulose at each side and spirally reticulate around the centre. The vertex is convex, narrowed behind and gradually widened out in front; it is marked by a faintly impressed median longitudinal line which extends from the base to the front where the junction of the vertex and the frons is very clearly demarcated by a deep and curved line which forms one side of an elliptical projection transversely placed between the eyes and jutting out so as to overarch the fronto-clypeus. The frons is hollowed and narrowed before the clypeus; centrally the hollow is marked with a network of lines in a characteristic spiral formation, and the puncturation of the raised sides adjoining the eyes is confluent with rugulose effect. Two obliquely-placed crescentic (fronto-clypeal) ridges, which extend to the clypeus, partly enclose the hollowed antennal areas at the lower or front end of the eyes. The clypeus is very short, greatly thickened, and widely emarginate in front in the middle, around the base of the labrum which is cordate and punctate with long yellowish covering hairs. The upper border of the narrow genae is a distinct ridge, bright golden, partly serrulate and surrounding the lower portion of each eye. The narrow and elliptical eyes are moderately prominent, placed vertically oblique and not close together; the finely faceted surface is golden amber with irregular longitudinal streaks of brown and scattered darker dots. The antennae are inserted in the large rounded hollows at a short distance from and on a level with the lower ends of the eyes. In this type specimen the segments of both antennae are wanting, except the long basal segment, which is lodged in the antennal groove, and the very short knob-like second segment. These antennal segments are dark metallic bronze green.

The *prothorax* is transverse, its breadth is nearly twice its length, it is broadest behind the middle where the sides bulge out a little, and it is somewhat angular and obliquely narrowed towards the front. The narrow front is margined by a thickened ring which is interrupted at the prothoracic episterna, and which is obsolete on the front of the pronotum about the middle and also close to the anterior angles.

The *pronotum* is a little sinuate in front; its base is markedly bisinuate, being excavated on each side of the median (scutellar) lobe for the reception of the elytral angulations. The sides of the pronotum (viewed from above) are sinuate; the side margins have the form of sinuate carinae which become inflected and hidden (when viewed from above) towards the forwardly projecting anterior angles. The pronotum is finely marginate and evenly convex, except at the front, directly opposite each eye, where the two thickened parts of the rim cause two shallow depressions. The surface of the pronotum is irregularly punctulate, except at the sides where the punctures are larger, oval-shaped and closer together, becoming confluent with rugulose effect within the anterior angles. The colour of the pronotum is dark metallic bronze green with irregular patches of violet on the outer or lateral portions.

The sides or pleura (*proepisterna* and *proepimera*) are broadly triangular, concave, reticulate and thinly punctulate, the punctules bearing fine golden hairs. The hollowed *mesepisterna* and the narrow and curved *mesepimera* are finely channelled. The moderately broad oblong *metepisterna* are in great part finely channelled and partly smooth with a few large punctures.

The *prosternum* is flattened and distally (behind the anterior coxal cavities) it is lobate at each side of the short

Buprestis dorsata Fab. × 8

triangular prosternal process which reaches the metasternum. The surface of the prosternum is finely punctured (punctulate), the punctures are irregular and not close together, and there is also a small slit-like depression at each side of the coxal margin. The *mesosternum* is punctulate. The *metasternum* is irregularly punctulate about the middle, the punctules bearing fine golden hairs; its lateral portions are finely channelled (a reticulate sculpturing with imbricate effect, the channels being a little raised). The ante-coxal portion of the metasternum is clearly marked off by an arcuate line, the ends of which almost reach the base, and it is divided by a median longitudinal suture which is slightly sulcate at the base and which extends from the base to the front of the metasternum.

The *scutellum* is roughly rhomboidal but posteriorly acuminate, and it is transversely carinate.

The *elytra* are unevenly convex, wider than the prothorax, and about twice as long as wide; the bases are angulate with sloping shoulders; the shoulder angles are sharply rounded and the shoulder callus is prominent; the sides are subparallel, very slightly sinuate, from the shoulder angles to about the middle and from there rather sharply rounded in and converging to the narrow apices. The sides are serrulate from beyond the middle to the sutural apices and the suture is marginate. The surface of the elytra is irregularly punctulate over the sutural area with crescentic punctures; on the sides the punctures are larger and round and closer together. There are three irregular depressions at the base, and three costae are faintly indicated, one near the suture, one about the middle and the third towards the outer side, and between these the elytral surface is irregularly depressed.

The front coxae are globular and sculptured on the surfaces with close and fine and branched channelling; the middle coxae also are globular but smaller and their surface is thinly punctulate; the hind coxae are transverse with their posterior margins oblique and slightly sinuate. The *legs* are short and clothed with fine hairs; the front femora are very stout (incrassate) and each has a conspicuous stout and sharp spine on the inner side; the femoral surface, which is sculptured with close and fine branched channelling, has a reticulate and imbricate appearance, and the channels are set with punctules bearing fine golden hairs. The middle and hind femora also are sculptured with imbricate effect but are less stout. The tibiae are punctate and set with raised ridges partly parallel and partly branched and vein-like. The front and middle tibiae are arcuate, but the hind tibiae are almost straight, slightly sinuate. The front tarsi have the first segment equal in length to the second and third together; the second tarsal segment has pointed lobes, the third is bluntly lobate, the fourth is very short with prolonged and blunt lobes, the fifth is very long and club-like, equal in length to the first four segments taken together, and the claws are simple. There are membranous pads beneath the first four tarsal segments which are punctate with strong hairs.

The first four sterna of the *abdomen* are irregularly and not closely punctulate with fine channelling on the lateral portions. The junction line or suture of the first two connate segments is distinct laterally but very faint (though clearly seen under the microscope) about the middle. The last or fifth abdominal sternum is the longest, its length is equal to the two preceding sterna taken together, it is widely and semicircularly emarginate, it has a strong median longitudinal carina, and its side margins, which are thickened and minutely reticulate, are bidentate at the tips. The surface of this sternum is marked with fine, branching, subparallel channelling (with punctures) which becomes obsolete anteriorly at each side of the carina, gradually passing into some crescentic punctures which bear fine golden hairs.

Length 14 mm.; breadth (across the shoulders of the elytra) 5 mm.
Hab. South America (Fab.).
See Plate 41.

A specimen in the Bishop Collection agrees closely with

the type, but its elytra are serrate rather than serrulate, the denticles are very irregular in size and a few of them are double-toothed.

75. *Chrysobothris quadrimaculata* (Fab.)

Coleopterorum Catalogus, pars 132 (J. Obenberger, 1934), Buprestidae, III, p. 648. *Catalogus Coleopterorum* (Gemminger and Harold, 1869), V, p. 1427. *Catalogus Buprestidarum* (Ed. Saunders, 1871), p. 100, No. 137. Wytsman, *Genera Insectorum*, Buprestidae (C. Kerremans), Fasc. XII, 1903, p. 192.

West Indies. Jamaica.

SYN. *Buprestis 4-maculata* Fab., *Gen. Ins.* p. 236, Nos. 31–32 (1776); *Sp. Ins.* I, p. 280, No. 46 (1781); *Mant. Ins.* I, p. 183, No. 69 (1787); *Ent. Syst.* I, 2, p. 209, No. 96 (1792); *Syst. Eleuth.* II, p. 208, No. 121 (1801); Oliv. *Ent.* II, 32, p. 76, pl. 10, fig. 110 (1790).

This species was founded by Fabricius on a specimen which belonged to Dr Fothergill. The example in Dr Hunter's Collection, in Cabinet B, drawer 8, under label

'*Bup. 4-maculata*
Fabr. pag. 280, No. 46'

is the insect described and figured by Olivier. The description given by Fabricius is very meagre; the elytral spots described by him as golden are, as stated by Olivier, coppery red, and the posterior spot is of a deeper hue than the anterior one.

Olivier's description is accurate as far as given; but it is very inadequate, and his illustration, though useful, is a crude portrayal of *quadrimaculata*. Fabricius mentioned India as the habitat, which was an error.

Description of Homotype, *Buprestis quadrimacu-*

lata **Fab.** Form elongate-ovate, broadest across the base of the elytra. The thorax subcylindrical, gently convex and punctulate in wavy formation with larger punctures at the sides of the disc. The elytra flattened, broadly rimmed and rather short, abruptly narrowed beyond the middle to the truncate apices, with angulate bases and slightly sinuate and angular outer sides serrulate towards the apex; and with transverse basal depressions, faint indications of costae, and closely punctulate. The anterior femora very stout and broad and each bearing a conspicuous short lateral spine.

The head metallic violet with amber golden eyes; the thorax cross-banded (fasciate) with golden green, deep violet and coppery red; each elytron with two irregular coppery red patches or spots placed lengthways, each patch or spot enclosed within a broad fascia of violet and separated by a transverse fascia of bright green confluent at each side with the green border around the elytron. The legs green and violet, and the under parts of the body dark green suffused with violet, except the outer parts of the posterior coxae which are brilliant golden. The thoracic sterna partly punctate, partly rugulose (prosternum) and finely channelled (mesopleura). The abdominal sterna irregularly and not closely punctulate. The punctures bearing yellowish hairs, long on the thoracic parts.

The short *head* is broad, it is as wide as the front of the thorax and metallic violet; it is unevenly but markedly flattened in front and its surface is closely punctate with irregular punctures and a little rugulose on the fronto-clypeal area, less closely punctate on the vertex. The vertex is narrow, it is impressed with coarse punctures and a distinct median longitudinal line which extends from the base to the front, where an arcuate thickening overarches the frons and marks off the vertex from the perpendicular front. The fronto-clypeus, which is narrowed between the insertions of the antennae, is unevenly flattened; it is coarsely and closely punctate and rugulose between and over the insertions of the

Buprestis quadrimaculata Fab. × 12

antennae, and the punctures bear long golden down-hanging hairs; the clypeus is short, widely emarginate, greatly thickened and outjutting over the base of the labrum, which is broadly lobate, convex and punctate with long covering hairs. The prominent reniform eyes are placed vertically oblique and near together, narrowing the vertex; the finely faceted corneal surface is amber golden with dark-coloured patches around the centre. The antennae, bronze green tinged with violet, are inserted in the small fronto-clypeal hollows, at a short distance from and on a level with the lower ends of the eyes. The first or basal antennal segment, which is partly lodged in the antennal groove, is the longest and stoutest of the series and is club-like; the short second and the long third segment are together about equal in length to the first; the succeeding eight serrate segments are short, about the same size as the second.

The *prothorax* is transverse, about twice as broad as long, it is broadest near the front where the sides are angulate and projecting. The narrowed front is bordered by a slightly thickened ring which is interrupted at the prothoracic episterna and which is obsolete on the front of the pronotum about the middle and also at the anterior angles.

The *pronotum* is slightly sinuate in front; its base is bisinuate, being excavated on each side of the median scutellar portion for the reception of the elytral angulations, and the sides (viewed from above) are angulate behind the narrowed front, projecting out and thence gently sinuate to the base. The pronotum is finely marginate in front, the narrow margin is continuous on the propleura and prosternum. The pronotal surface is evenly convex, except at the front, opposite each eye, where there are two oblique linear depressions marking off portions of the slightly thickened rim, and also at the middle of the base, where there is a shallow cordate depression; otherwise the surface is irregularly but not closely punctulate, except about the front angles where it is punctate with larger punctures.

The colour pattern of the pronotum is, like that of the elytra, very distinctive and in the form of alternate irregular cross-bands or fasciae. The front fascia is golden green, it narrows towards the front angles; behind it there is a broader fascia of deep violet and a narrower one of the same colour which extends across the base. Between these violet bands there is a broad fascia of coppery red.

The sides or *pleura* (*proepimera* and *proepisterna*) are broadly triangular, concave and punctulate with fine golden hairs. The bronze green *mesepimera* and the hollowed *mesepisterna* are very finely channelled. The broad, oblong and concave *metepisterna* are violet coloured and partly punctate. The *prosternum* is punctate, rugulose anteriorly, flattened and violet coloured and the narrow prosternal process reaches the metasternum. The *mesosternum* is bright bronze green and punctulate; it has the form of two convexities, with the prosternal process lodged between these. The *metasternum* is metallic bronze green and punctulate about the middle and its lateral portions are violet coloured and closely punctate. The ante-coxal portion of the metasternum is marked off by an arcuate line, the ends of which nearly reach the base, and it is divided by a median longitudinal sulcus; its forward continuation is a clearly impressed line which extends to the front of the metasternum. The punctures of the thoracic sterna bear long golden hairs.

The *scutellum* is destroyed in this type, the specimen having been pinned through that part.

The *elytra*, a little wider than the prothorax and nearly twice as long as wide, are flattened rather than convex, for the convexity is slight; the surface is uneven, and each elytron has a rim-like border extending from the shoulder along the outer side and about halfway along the suture from the apex. The bases are angulate with sloping shoulders; the shoulder angles are rounded and the shoulder callus is prominent; the sides are subparallel, very slightly sinuate, for nearly two-thirds of their length, then sharply rounded in and converging straightly to the narrow truncate and tridenticulate apices. The sides are serrulate from the middle to the sutural apices and the suture is very finely marginate towards the sutural denticle.

The surface of the elytra is marked with irregular depressions and is closely and irregularly punctulate; on each elytron there is a double fovea at the base and two wavy costae are faintly indicated, one subsutural and the other near the middle. Each elytron is bordered all round with metallic green, and confluent with this border is a transverse fascia of the same colour situated beyond the middle, nearer to the apex than the base. The green border and the fascia are edged with gold. Above the fascia is a large light coppery red patch or spot surrounded with deep violet, and below the fascia, i.e. within the apical area, there is a lesser patch of deep coppery red within a triangle of deep violet colour.

The *legs* are short, dark green with violet and thinly punctate with long hairs. The front and middle coxae are globular and about the same size; the surface of the former is pitted with coarse punctures and partly marked with fine channelling and the surface of the latter is punctulate. The hind coxae are transverse, with the posterior margin oblique and sinuate, and the coloration is green and violet with the outer portions brilliant golden. The front femora are very stout (incrassate), stouter than the middle and hind femora, and on the inner side of each femur there is a conspicuous short stout spine. The trochanters of the middle legs are bright golden. The front and middle tibiae are arcuate, the hind tibiae are straight. The hind tarsi are violet coloured; the first and second tarsal segments are about equal in length, the third is smaller, the fourth is very small with long lobes and the fifth or terminal segment is long, narrow and clavate with the thickened end terminal. The length of the fifth tarsal segment is almost equal to that of the other segments taken together, and it bears two simple claws.

The first four sterna of the *abdomen* are set with long and fine golden hairs and are irregularly but not closely punctulate, except the lateral parts of the first sternum which are closely pitted. The suture of the first two connate segments is very faint but clearly seen (under the microscope) about the middle. The last or fifth abdominal sternum, which is the longest, is equal in length to the two preceding sterna taken together; its thickened side margins are bidentate at the tips, between which it is widely emarginate and with a small median lobe; it is not carinate and its surface is thinly punctate. The coloration of the abdominal sterna is dark metallic green suffused with deep violet, with bright green on the middle portions of the posterior borders.

Length 10 mm.; breadth (across the shoulders of the elytra) 4 mm.

Hab. India (Fab. and Oliv.).

See Plate 42.

76. *Belionota canaliculata* (Fab.)

Coleopterorum Catalogus, pars 132 (J. Obenberger, 1934), Buprestidae, III, p. 672.

Tropical and South Africa.

SYN. *Buprestis canaliculata* Fab., *Mant. Ins.* I, p. 181, No. 58 (1787); *Ent. Syst.* I, 2, p. 205, No. 82 (1792); *Syst. Eleuth.* II, p. 204, No. 102 (1801). *Catalogus Coleopterorum* (Gemminger and Harold, 1869), V, p. 1420. *Catalogus Buprestidarum* (Ed. Saunders, 1871), p. 92, No. 14. Wytsman, *Genera Insectorum*, Buprestidae (C. Kerremans), Fasc. XII, 1903, p. 197.

Belionota championi Murray, *Trans. Linn. Soc. London*, XXIII, 1869, p. 451, pl. 47, fig. 5. *Catalogus Buprestidarum* (Ed. Saunders, 1871), p. 92, No. 14.

As stated by Fabricius, the type of this species is in the Banks Collection, British Museum. The Hunterian example in Cabinet B, drawer 8, under label

'*Bup. canaliculata*
Fabr. MSS'

has been compared with the type and with modern examples in the British Museum, and it closely corresponds. Incidentally this specimen is one of the best preserved Buprestids in Dr Hunter's Collection, the antennae, limbs and other parts being remarkably complete.

77. *Coraebus meditabundus* (Fab.)

Coleopterorum Catalogus, pars 143 (J. Obenberger, 1935), Buprestidae, IV, p. 831. Théry, *Ann. Soc. Ent. France*, XCVI, 1927, pp. 257–258.

Assam.

SYN. *Buprestis meditabunda* Fab., *Mant. Ins.* I, p. 183, No. 80 (1787); *Ent. Syst.* I, 2, p. 212, No. 112 (1792); *Syst. Eleuth.* II, p. 211, No. 138 (1801); Oliv. *Ent.* II, 32, p. 74, pl. 10, fig. 107 (1790).

Cisseis meditabunda (Fab.), *Catalogus Buprestidarum* (Ed. Saunders, 1871), p. 103, No. 25. Wytsman, *Genera Insectorum*, Buprestidae (C. Kerremans), Fasc. XII, 1903, p. 229.

The single example in Cabinet B, drawer 8, under label

'*Bup. meditabunda*
Fabr. MSS'

is evidently the type, it answers the short descriptions given by Fabricius and Olivier and it is clearly the insect figured by Olivier.

Description of Type, *Buprestis meditabunda* Fab. Form elongate-oblong, rather narrow, subparallel from the shoulders and convergent apically; a little constricted about the middle of the elytra and broadest behind the middle, being there broader than at the base. The prothorax markedly convex with a broadly lobate bisinuate base and with strong lateral carinae which slope obliquely forwards and downwards, and closely punctulate over the disc. The elytra narrowly flattened along each side of the raised suture, the outer portions convex with prominent shoulder callosities, broadly triangular basal foveae and with slightly depressed areas at regular intervals and very finely pitted; the elytral surface closely punctulate and honeycomb-like, and (amidst the rounded punctules) very fine pits on the edges of the comb; the borders marginate, the sutural margins elevated, the entirely serrulate outer margins obsolete at the middle of the base and apically; the elytral apices entire, rounded in, and a little divergent at the suture. The head is metallic bronze suffused with violet and with light yellowish green eyes; the prothorax is metallic green with tinges of violet, except about the flattened angles, which are bright golden green; the elytra are green strongly suffused with violet. The under parts of the body are brilliant metallic green, the mouth-parts deep violet. The prosternum is closely punctate, the mesosternum is not visible and the metasternum is less closely punctulate; on the thoracic pleura the punctules become confluent, forming fine channels. The abdominal sterna are irregularly punctulate and finely channelled along the sides, where the punctules become confluent. All the limbs are wanting in this type specimen.

The short and broad *head* is mainly metallic bronze violet and its surface is closely punctate and rugulose. The vertex is marked by a slightly raised and very fine median longitudinal line; this line becomes obsolete in the deep and broad median longitudinal sulcus or furrow of the perpendicular frons, which has a thickened anterior border. The short and widely emarginate clypeus, distinctly marked off from the frons, is narrowed between the antennal insertions. The clypeus and the labrum are deep violet coloured and rugose. The prominent eyes are oval, vertically placed and wide apart; the finely faceted corneal surface is light yellowish green with a narrow dark-coloured marginal border and with dark patches around the centre. The antennae are inserted in large oval excavations of the clypeus, which are directed obliquely towards the lower ends of

Buprestis meditabunda Fab. × 9

the eyes and which there open as short grooves on the gena; the first or basal segments of the antennae, which are partly lodged in the short grooves, are club-shaped and deep violet coloured. The other segments of both antennae are wanting.

The *prothorax* is transverse, nearly twice as broad as long, its breadth being 3½ mm. and its length 2 mm.; it is broadest at the base, where its outline is almost continuous with that of the elytra. The front is narrowed, its width is 1 mm. less than that of the base, and it is subcylindrical. The *pronotum* is slightly sinuate in front and slightly but widely arcuate about the middle; the base is bisinuate, with a wide median scutellar lobe and two angular excavations or bays for the reception of the obtusely angulate elytral projections. The sides of the pronotum (viewed from above) have the form of expanded rims, which are slightly serrulate along their outer edges and which curve gently outwards from the sharply right-angled posterior angles and then gradually inwards towards the head, thus narrowing the front. These rims form the lateral carinae which slope obliquely forwards and downwards to the gena at the lower end of each eye, there meeting the front edge of the pronotum, each side, where it ends or becomes continuous with the sinuate border of the propisternum. The pronotum is plain-edged in front, thinly fringed with short and fine golden hairs and narrowly marginate on the sides and the base.

The surface of the pronotum is very convex from the front to the middle; beyond the middle, i.e. on the scutellar area of the disc, it becomes flattened. On each side of the scutellar area the convex disc is interrupted by an oblique furrow, and thus two small portions of the disc are almost isolated. These small elevated parts surmount the two angular excavations of the bisinuate base and, in close conjunction with similar elevations of the angulate projections of the elytra, they form two small prominent basal convexities. The aforementioned furrows, which become obsolete at the scutellar lobe, are oblique prolongations of the lateral portions of the pronotum. As the convex disc is curved in towards the base and therefore narrowed behind, the lateral portions of the pronotum bordering the disc each side represent areas which are of considerable extent, roughly triangular and irregularly flattened.

The surface of the disc is metallic green suffused with violet and is closely and irregularly punctulate, except at the middle of the front, where it is golden green and a little rugulose. The lateral portions are golden green and the irregular surface is marked with oval puncturation less close but tending to confluence in lines radiating out from the posterior angles. The narrow lateral margins are defined each by a marginal line of confluent punctures parallel with the serrulate outer edge; this line begins at the posterior angle, and there alongside it is a short but deep linear impression.

The sides or *pleura* (*proepisterna* and *proepimera*) are broadly triangular and bright metallic green; the surface is vermiculate with confluent puncturation, and the proepimera are set with short yellowish hairs. The narrow *mesepimera* and *mesepisterna* are bright metallic green, vermiculate and closely set with short golden hairs. The narrow, oblong and convex *metepisterna* are bright metallic green and vermiculate but with some separate punctures anteriorly.

The *prosternum* is closely punctate and bright metallic green; its middle portion is flattened and the flat prosternal process, which is relatively broad and dagger-shaped, reaches the metasternum, the blunt tip of the process being lodged in the notched front of the metasternum. The lateral parts of the *mesosternum* are very short and set back on the sides, very inconspicuous, and rugose. The *metasternum* is bright metallic green, it is irregularly and not closely punctulate about the middle; its front portion and its convex lateral portions are vermiculate with confluent puncturation. The ante-coxal portion of the metasternum is marked off by an arcuate wavy line, the ends of which nearly reach the base, and it is divided by a clearly impressed median longitudinal line which extends to the front of the metasternum and which becomes sulcate about the middle.

The *scutellum* is large, flattened, mainly smooth, bright golden green and of a characteristic shape like that of a brace bracket; it is triangular, with a sharp apex but with the sides deeply incurved except about the base which is transverse with outjutting rounded ends.

The *elytra*, metallic bronze green with a strong suffusion of violet about the base, are of the narrow form; the length is more than twice the width; the width across the base is the same as that of the base of the pronotum, but behind the middle the width is greater than across the base. The base of each elytron has an obtusely angulate raised projection, about the middle, which is lodged in the corresponding angular excavation of the pronotal base. The bases are marginate, except between the scutellum and the apex of the angulate projection, where the narrow and plain margin is replaced by a line of oval punctures. The shoulder angles are slightly obtuse, the sides being there rounded in. The sides are slightly sinuate (subparallel) from the base to the middle and very distinctly rounded out beyond the middle, thence straightly convergent to the narrowed and entire apices, which are obliquely rounded in with the extreme tips slightly protuberant at the suture and there a little divergent. The sides are marginate, except on the oblique apices, and the margins are finely serrulate. The suture is marginate on three-quarters of its length from the apex, and the sutural margins are elevated. The elytra are narrowly flattened along each side of the raised suture and convex on the outer portions; the shoulders form prominent callosities and along with the elevations of the basal angular projections, which are a little rugulose, partly enclose the roughly triangular basal fovea on each elytron. Between the basal foveae and behind the scutellum there is a wide but shallower median depression involving the area on each side of the anterior and non-elevated portion of the suture.

The surface of the elytra is irregularly and closely punctulate; anteriorly the punctures are rounded, towards the middle they are somewhat oval, posteriorly they become crescentic and confluent here and there, and on the apices they show linear regularity. Thinly scattered amidst these punctules there are round ones of minute size.

Upon each elytron there are six small and slightly depressed spot-like areas, which are closely and finely pitted. Four of these pitted spots are situated in a row, at regular intervals apart, alongside and near the suture between the median depression and the apex; the other two are near the outer margin and are not directly but obliquely opposite the second and third subsutural spots. The first subsutural spot is round, the others are more oval; the fourth subsutural spot is the broadest and would appear to be a confluence of two due to the narrowing of the apex.

The globular front coxae are larger than the middle coxae and the trochantins of both have finely milled surfaces. The hind coxae are transverse and concave, very slightly dilated at the inner side, and the posterior margins are horizontal; the inner half of the bright metallic green surface is sparsely punctulate, the outer half is vermiculate, and there are a few scattered golden hairs. The other parts of the *legs* are wanting.

The sterna of the *abdomen* are uniformly brilliant metallic green; the first sternum is the longest, the second and fifth are shorter than the first and about the same length, the third and fourth sterna are the shortest and are of equal length. The first four abdominal sterna are irregularly punctulate; the punctules are thinly scattered on the middle parts and are close together on the sides, where there is considerable confluence producing a vermiculate appearance of broken and irregularly waved lines or channels. The junction line of the connate first and second segments is faint but clearly seen (under the microscope) about the middle and also towards the outer edges. The last or fifth sternum, which is one and a half times the length of the preceding one, is terminally rounded and its surface is vermiculate with irregular fine channels and without free punctures.

Length 12 mm.; breadth (across the elytra behind the middle) 4¼ mm.

Hab. North America (Fab. and Oliv.).

See Plate 43.

78. *Agrilus ruficollis* (Fab.)

Coleopterorum Catalogus, pars 152 (J. Obenberger, 1936), Buprestidae, v, p. 1224. *Catalogus Coleopterorum* (Gemminger and Harold, 1869), v, p. 1445. *Catalogus Buprestidarum* (Ed. Saunders, 1871), p. 116, No. 101. Wytsman, *Genera Insectorum*, Buprestidae (C. Kerremans), Fasc. XII, 1903, p. 270. *Catalogue of the Coleoptera of America, North of Mexico* (Charles W. Leng, 1920), p. 184. *The Buprestidae of Pennsylvania* (Joseph N. Knull, 1925), p. 39 (The Ohio State University Studies, Vol. II, No. 2, December 1925), etc.

U.S.A. and Canada.

SYN. *Buprestis ruficollis* Fab., *Mant. Ins.* I, p. 184, No. 85 (1787); *Ent. Syst.* I, 2, p. 214, No. 121 (1792); *Syst. Eleuth.* II, p. 213, No. 152 (1801); Oliv. *Ent.* II, 32, p. 78, pl. 9, fig. 101 (1790).

In Cabinet B, drawer 8, under label

'*Bup. ruficollis*
Fabr. MSS'

there are two specimens (co-types) of this species. One of these co-types has been compared with a modern example of *ruficollis* Fab. in the British Museum Collection and it shows exact similarity in specific features. Both co-types answer the descriptions given by Fabricius and Olivier, and Olivier's figure, drawn the natural size but not very clearly coloured, obviously resembles them.

Description of Co-type, *Buprestis ruficollis* Fab. Form narrow and slender, elongate-oblong, subparallel, rounded in outline at the sides of the prothorax and sinuate along the sides of the elytra, which are constricted about the middle and narrowed towards the rounded and serrulate apices. The head, including the prominent eyes, nearly as broad as the front of the prothorax, which (across the middle) is as broad as the base of the elytra; the width of the elytra between the middle and the apices greater than that of the prothorax. The prothorax subcylindrical and obliquely carinate on the sides. The pronotum markedly convex in front, deeply hollowed laterally, less so behind; the base nearly straight in the middle but sinuate to each angle, and the sides (viewed from above) arcuate. The elytra hollowed behind the sinuate bases, the sutural portions flattened, the lateral portions moderately convex, and the apices tumid. The surface punctulate and either rugulose or shagreened on parts; and the coloration metallic bronze green, dull and very dark on the elytra, brilliant brassy about the eyes and coppery on the head and the pronotum.

The short and insunk *head* is coppery bronze green with a bright brassy patch above each eye. About the vertex the surface is rugulose; on the perpendicular front it has a more imbricate sculpture with linear punctules bearing short yellowish-white hairs. On the vertex there is a faintly impressed median longitudinal line; it intersects the deep, broad and rounded median longitudinal sulcus on the frons and becomes obsolete towards the middle of that region where the furrow or sulcus ends. The frons is abruptly narrowed between the insertions of the antennae, and is there distinctly marked off from the narrow and roughly hemispherical clypeus by a thin arcuate ridge. The clypeus is widely emarginate before the lobate labrum. The prominent oval eyes are vertically placed and wide apart; the finely faceted corneal surface has a bright golden margin around a darker coloured centre and is surrounded by a narrow and dark-coloured insunk border. The short antennae are serrate, except the first three segments and the

Buprestis ruficollis Fab. × 16

last or eleventh, which are club-shaped; the first or basal segment is the largest, the second and the third are progressively smaller. The antennae are inserted in large pyriform excavations of the clypeus, which are obliquely directed towards the lower ends of the eyes, where the narrow part of the pear-shaped socket is open at the narrow gena.

The *prothorax* is subcylindrical and its sides are strongly carinate; the carinae or keels are sinuate and are directed obliquely downwards and forwards from the posterior pronotal angles to the gena at the lower end of each eye. The *pronotum* is transverse, its breadth is one-half greater than its length which measures 1 mm.; it is broadest across the middle and the base is a little narrower than the front. The front of the pronotum is sinuate, it is widely arcuate between the eyes; the base is bisinuate, but the middle or scutellar part of the base is almost straight and the two sinuations or bays are wide rather than deep; the hind angles are sharp. The sides of the pronotum (viewed from above) are hollowed about the middle and have arcuate rims, these rims being the lateral carinae already described; as the carinae are obliquely placed, the sides are therefore steeply deflected in front and the sharp front angles are very low down at the gena, near the lower ends of the eyes. The disc of the pronotum is not entirely raised; it is markedly convex from the front margin to about the middle and also around the two basal sinuations or bays, but the area between these convex parts is furrowed (sulcate) and continuous with the deeper hollow on each side. There is also a slight central fovea on the main convexity. The pronotum is finely marginate in front and behind; its surface is transversely and sinuously rugulose, with minute linear punctures at regular intervals between the wavy wrinkles, and the coloration is coppery red.

The sides or *pleura* (*proepisterna* and *proepimera*) are somewhat semicircular and centrally convex, and the surface is metallic bronze green and irregularly punctulate, the punctules somewhat triangular and bearing short and fine yellowish hairs. The surfaces of the narrow *mesepisterna* and *mesepimera* and of the narrow, oblong and slightly convex *metepisterna* (somewhat vertically placed) are similar to those of the propleura as described above.

The *prosternum* is metallic bronze green, roughly triangular and lightly rugulose, most of the punctures are confluent in regular transverse lines and bear short and fine hairs; the main portion is gently rounded and the prosternal process, which is lobe-shaped and convex, reaches the metasternum, the blunt tip of the process being lodged in the notched front of the metasternum. The lateral parts of the *mesosternum* are set back on the sides and scarcely visible. The *metasternum* is metallic bronze green and irregularly and rather closely punctulate, especially on the concave sides, which have a rugulose appearance; the punctules are variable and bear short yellowish hairs. The middle portion of the metasternum is rounded, there is a short median sulcate depression which narrows towards the base, the surface is punctulate, and the fine angular punctures are connected together with imbricate effect; the ante-coxal portion is partially marked off by two short and curved impressed lines which reach the anterior borders of the hind coxae.

The *scutellum* is dull bronzed black; it is large and triangular, but the sides are deeply incurved, and the transverse base is distinctly sunk and marked off from the apical portion by a strong transverse carina.

The *elytra* are dull bronzed black and very narrow in form, the length (5 mm.) being three times greater than the average width. Across the base the width is equal to that of the pronotum across the middle, but between the middle and the apices it is a little more. The base of each elytron is sinuous, it is widely rounded off against the scutellum and about the middle there is a small raised projection, rounded rather than angulate, which is lodged in the corresponding bay of the pronotal base. The elytral bases are finely marginate and the shoulder angles are obtuse and rounded out at the sides. The sides of the elytra are subparallel from the base to the apical third; and they are sinuate, slightly rounded out behind the shoulder angles, gradually constricted towards the middle and distinctly rounded out or dilated from the middle towards the narrowed apices, which are entire with the tips regularly rounded (arcuate) in outline and noticeably tumid. The sides are also finely marginate as far as the apical portion, where the margin becomes obsolete and is replaced by fine denticulation most clearly seen on the serrulate tips. The suture is distinctly marginate on three-quarters of its length from the apex to beyond the middle.

The shoulders form prominent callosities and, along with the sinuate base, partly enclose the deep hollow (basal fovea) on each elytron; the sutural portions of the surface are flattened, the lateral portions are moderately convex, and over the apices the surface is raised (tumid). The elytral surface is asperate with shagreened sculpture.

The globular front and middle coxae are about the same size; the hind coxae are transverse and partly concave, a little expanded at the inner side, the posterior margins are horizontal and sinuate, the bronze green surface is closely punctulate with rugulose effect and the punctules bear very short fine hairs.

The short *legs* are uniformly glossy dark bronze green, but with bright coppery touches on the tibiae, and the surface is puncturate, the punctures bearing very short yellowish hairs. The femora are moderately stout; the first segment of the five-segmented hind tarsi is as long as the second, third and fourth taken together. The tarsal segments have small membranous pulvilli and the claws are apparently cleft, with the inner portion shorter and directed inwards.

The sterna of the *abdomen* are uniformly metallic bronze green. The first sternum is the longest, the second appears to be about as long as the fifth, the third is a little shorter, and the fourth is the shortest. The connate first and second segments are so completely fused that only a short raised overlap on each side indicates the posterior border of the first sternum; and there is a small but distinct round fovea centrally placed in the middle line of these two conjoined segments. The fifth sternum is arcuate with the rounded posterior border pitted. The surface of the abdominal sterna is finely punctulate and has an imbricate appearance, most of the punctules being angular and connected and they have very short and fine yellowish hairs.

This co-type is apparently a female, the first two abdominal sterna being without a groove.

Length 7 mm.; breadth (across the elytra, beyond the middle) 1¾ mm.

Hab. America (Fab. and Oliv.).

See Plate 44.

79. *Agrilus biguttatus* (Fab.)

Coleopterorum Catalogus, pars 152 (J. Obenberger, 1936), Buprestidae, v, p. 961. *Catalogus Coleopterorum* (Gemminger and Harold, 1869), v, p. 1437. *Catalogus Buprestidarum* (Ed. Saunders, 1871), p. 118, No. 152. Wytsman, *Genera Insectorum*, Buprestidae (C. Kerremans), Fasc. XII, 1903, p. 274.

Europe, Algeria, Transcaucasia, Asia Minor, Persia.

SYN. *Buprestis biguttata* Fab., *Gen. Ins.* p. 237, Nos. 39–40 (1776); *Sp. Ins.* I, p. 281, No. 55 (1781); *Mant. Ins.* I, p. 184, No. 82 (1787); *Ent. Syst.* I, 2, p. 213, No. 115 (1792); *Syst. Eleuth.* II, p. 212, No. 144 (1801); Oliv. *Ent.* II, 32, p. 76, pl. 7, fig. 75 (1790).

Fabricius founded this species on a specimen in the Yeats Collection. Dr K. G. Blair suggested the possibility of this insect having been acquired by Dr Hunter. In Cabinet B, drawer 8, there are two specimens under label

'*Bup. biguttata*
Fabr. pag. 281, No. 55'

which answer the descriptions of this species given by Fabricius and Olivier and correspond to Olivier's figure of *biguttata* Fab.; the larger one is a male (with the genitalia extruded), the smaller one is a female. They have been compared with examples in the British Museum Collection and appear to be normal British specimens. There is, however, no definite evidence that either of these beetles originally belonged to the Yeats Collection; they may, however, be metatypes.

Description of Metatype, *Buprestis biguttata* Fab.

Female. Form narrow and elongate, subparallel between the head and the narrowed apical third of the elytra, rounded in outline at the sides of the prothorax and somewhat sinuate along the sides of the elytra, which are a little constricted about the middle, gently rounded out behind the middle and from there straightly convergent towards the regularly rounded (arcuate) apices which are slightly expanded, entire, and finely serrulate with a short gap near the outcurved ends of the sutural margins. Markedly flattened above and convex beneath.

The front of the head flattened and with a central depression. The prothorax not quite as broad as the elytra, cylindrical, transverse, and obliquely carinate on the sides, the sinuous carinae (side rims of pronotum) extending from the posterior angles obliquely downwards to the deflected sharp anterior angles. The pronotum slightly convex across the front, centre and base, and furrowed between these parts; almost straight in front and the base bisinuate. The scutellum with a ridge across the base. The elytra long and narrow, at the base slightly broader than the thorax, and the sides sharply convergent from the middle to the apices; flattened over the disc, convex along the sides, and with two incurved basal hollows between the prominent long shoulders and the flattened scutellar area; the elytral surface more shagreened than rugulose in appearance, and with two small white-haired spots adsutural and towards the apex. Abdomen with six white-haired spots ventrally. Coloration bright metallic bronze green with localised suffusion of violet. The undersurface finely and more or less closely punctate with considerable confluence of the punctures on the thoracic parts.

The short *head* is brilliant bronze green and rugulose-punctate. Upon the vertex there is an impressed median longitudinal line which is continuous on the perpendicular front; it intersects the median longitudinal sulcus of the frons and becomes obsolete where the sulcus forms a broad frontal depression or fovea. A straight line between the uppermost parts of the antennal sockets marks off the frons from the narrow clypeus which is a narrow subquadrate area between the antennal sockets and which is widely emarginate before the small lobate labrum. The prominent oval eyes are vertically placed and wide apart; their finely faceted corneal surfaces are dull green with darker patches and some small golden spots. The short antennae are metallic dark bronze green and are serrate, except the first three segments and the last or eleventh which are club-shaped; the first or basal segment is the largest, the second and third are about equal in size. The antennae are inserted in large pyriform excavations between the clypeus and the lower portions of the eyes, the sockets being directed obliquely towards the lower ends of the eyes, where the narrow part of the pear-shaped socket is open at the narrow gena.

The *prothorax* is cylindrical and transverse and its sides are strongly carinate; the sinuate carinae are directed obliquely downwards and forwards from the posterior angles of the pronotum to the front margin where it meets the pro-episternal borders, and there join the margin to form the deflected sharp anterior angles, behind the eyes, at a short distance from their lower ends. The *pronotum* is transverse, its breadth (2¼ mm.) is greater than its length, which measures 1½ mm.; it is broadest across the middle and the base is slightly narrower than the front. The front of the pronotum is almost straight, very slightly arcuate between the eyes; the base is bisinuate, the middle or scutellar part of

Buprestis biguttata Fab. ♀ × 9

the base is broad and straight and the two sinuations or bays are wide but not deep, widely angular; the posterior angles are sharp and obtuse angles. The sides of the pronotum (viewed from above) are rounded and have arcuate narrow rims (the lateral carinae); as the carinae are obliquely placed, the sides are considerably deflected in front and the sharp anterior angles are low down at the gena and at a short distance from the lower ends of the eyes. The disc of the pronotum is not entirely raised, it is a little convex at the front and across the centre and also around the basal sinuations or bays; between these convex parts the disc is furrowed (sulcate), and the sides of the disc are somewhat hollowed about the middle. The pronotum is finely marginate all round, the front portion of the margin being brilliant and most apparent. The surface of the pronotum is transversely and sinuously rugulose, and on the wrinkles there are minute round punctules at intervals apart; there are also some larger punctures in front and within the posterior angles. The general appearance of the surface is that of a shagreened sculpture with slightly rugulose effect. The coloration of the pronotum is metallic bronze green suffused with violet.

The sides or *pleura* (*proepisterna* and *proepimera*) are broadly angular, markedly convex beneath and involving the front portion of the lateral carinae, and a little hollowed about the middle; and the dark bronze green surface is closely puncturate and finely rugulose, the punctures bearing very short and fine whitish hairs. The colour and sculpture of the short and narrow *mesepisterna* and *mesepimera* and of the narrow and oblong *metepisterna* (somewhat vertically placed) appears to be the same as that of the propleura.

The *prosternum*, dark bronze green, is roughly triangular; its base is emarginate and sinuous and is marked off from the middle portion, by a deep transverse furrow, as a distinct gular part. The base and the middle portion of the prosternum are moderately convex; and the prosternal process, which is lobe-shaped and flattened, reaches the metasternum, the blunt tip of the process being lodged in the emarginate front part of the metasternum. The prosternal surface is irregularly and closely punctate, and many of the punctures are confluent. The lateral parts of the *mesosternum* are not clearly visible. The *metasternum* is dark bronze green and irregularly punctate, with very short and fine whitish hairs; but most of the punctures are confluent in broken lines. The surface is convex, except over the ante-coxal area, where it is flattened and depressed. A deeply impressed median longitudinal line extends from the deep angular notch on the base towards the front, becoming obsolete near the frontal emargination. This line is crossed near the middle by a transverse line which bends abruptly at each side and extends obliquely to the edge of the base, thus reaching the anterior borders of the hind coxae and marking off very clearly the ante-coxal pieces. Where the two lines cross, the metasternum is distinctly hollowed.

The *scutellum* is bright metallic bronze green tinged with violet; it is large and triangular, but the sides are rounded at the base and are deeply incurved towards the apex and upon the large transverse base there is a strong median transverse ridge or carina. In front of the ridge the surface is strigose, behind the ridge it is asperate.

The *elytra* are bright metallic bronze green with violet along the narrowly deflected sides and upon the apices. Between the middle and the apices, and close together at the sutural margins, there are two small and irregular white spots, these being slight depressions with overlying silvery white recumbent hairs of considerable length. The length of the elytra (8½ mm.) is more than three times the breadth (2½ mm. across the shoulders), which is slightly greater than that of the pronotum across the middle. The elytral base is rather thickly marginate and that of each elytron is sinuous, widely rounded off against the scutellum and having about the middle a wide arcuate projection which is received in the corresponding bay of the pronotal base. The sides, subparallel from the base to the apical third, are sinuate, rounded out behind the shoulder angles, gradually but slightly constricted behind the shoulders to about the middle, gently rounded out behind the middle and from there straightly convergent towards the regularly rounded (arcuate) apices, which are distinctly expanded and entire and finely serrulate with a short gap near the outcurved apical ends of the sutural margins. The sides are also finely marginate with minute denticulation as far as the apical third; there the margin becomes obsolete and is replaced by the denticulation, which is continuous and strongest around the serrulate tips. The suture is finely marginate from the middle of the apices to a point within a short distance from the scutellum, where it becomes obsolete or is hidden from view. The shoulder angles are rounded, the shoulders are prominent and long; and on each elytron, behind the sinuous base and between the shoulders and the flattened scutellar area, there is an incurved hollow (basal fovea). The elytral surface is flattened above, a little hollowed along each side of the suture from the middle towards the apex where the sutural margins are raised, and it is narrowly convex along the sides with marked epipleural deflexion of the shoulder region. The surface is punctate and more shagreened than rugulose in appearance.

The short *legs* are uniformly dark metallic bronze green, brighter on the tibiae, and the leg surface is finely punctulate with very short and fine whitish hairs.

The front and middle coxae are globular and about equal in size; the hind coxae are transverse and a little concave, the inner portions are considerably expanded, the posterior margins are widely arcuate, and the sculpture of the surface is similar to that of the metasternum.

The femora are fairly stout; the front and middle tibiae are curved, and the hind tibiae are almost straight. The first segment of the five-segmented hind tarsi is as long as the second, third and fourth taken together. The tarsal segments have small membranous ventral pulvilli; the claws are cleft and their inner portions are shorter, directed inwards and contiguous.

The *abdomen* is uniformly dark metallic bronze green with a strong suffusion of violet. The proximal sternum (first and second sterna conjoined) is very long, very nearly half the length of the abdomen. The connate first and second abdominal segments have been so completely fused that only a short and rather faint transverse indentation on each side indicates the posterior border of the first sternum. The third sternum is longer than the fourth, which is the shortest, the fifth is one and a half times the length of the third, it is arcuate and its rounded margin is narrowly and evenly bordered by an impressed line. The surface of the abdominal sterna is finely and not closely punctulate, except about the tip of the fifth sternum where larger punctures occur close together. There are six white spots (slight depressions of the surface covered with moderately long overlying silvery-white hairs) on the third, fourth and fifth sterna, one pair on each and antero-lateral in position.

Length 10½ mm.; breadth (across the shoulders of the elytra) 2½ mm.

Hab. England (Fab.), France, England (Oliv.).

See Plate 45.

The other specimen is a *male* with the genitalia extruded; its length is 11½ mm. The anterior tibiae have a small sharp hook at the distal end on the *inner* side. The suffusion of violet on the elytra is more extensive than in the female metatype.

Family ELATERIDAE

Sub-family HEMIRRHIPINAE

80. *Hemirrhipus fascicularis* (Fab.)

Coleopterorum Catalogus, pars 80 (S. Schenkling, 1925), Elateridae, I, p. 52. *Catalogue of the Coleoptera of America, North of Mexico* (Charles W. Leng, 1920), p. 167.

Brazil up to United States.

SYN. *Elater fascicularis* Fab., *Mant. Ins.* I, p. 171, No. 2 (1787); *Ent. Syst.* I, 2, p. 216, No. 2 (1792); *Syst. Eleuth.* II, p. 222, No. 3 (1801); Oliv. *Ent.* II, 31, p. 8, pl. 5, fig. 56 (1790).

The single example in Cabinet B, drawer 7, under label

'*El. fascicularis*
Fabr. MSS'

is apparently the type, it answers the descriptions given by Fabricius and Olivier and corresponds to Olivier's illustration; it has been compared with specimens of this species in the British Museum Collection and found to agree. The modern examples of *fascicularis* Fab., from Venezuela, in the Bishop Collection very closely resemble this type.

Description of Type, *Elater fascicularis* Fab. Form elongate, oblong ovate, ellipsoidal transversely, the body outline (viewed from above) regularly continuous and (side view) with a slight elytral hump; convex above, but less so over the elytra towards the apex and deeply recessed around the insunk scutellum. The head chestnut brown with a large frontal recess, the pronotum glossy dark chestnut brown. The elytra pale yellowish brown with two reddish brown markings at the base, chestnut brown marking in the form of a fascicular pattern about the middle and a narrow wavy transverse band near the brown-coloured apex; and striate-punctate, with fine punctulation on the spaces between the punctured striae. Eyes prominent, round, convex, and shining blackish brown; the antennae dark chestnut brown and pectinate; the pygidium partly exposed. The entire dorsal surface coated with fairly long, fine, pale brown and recumbent hairs. Undersurface uniformly dull glossy dark chestnut brown, punctate on the thoracic parts, finely punctulate on the abdomen, and lightly covered with pale brown recumbent hairs; the legs hairy and of a lighter hue.

The short insunk *head* is chestnut brown, irregularly and closely punctate and covered with yellowish white hairs; it is almost square in front but sharply angled off towards the clypeus. The frons is deeply hollowed (frontal recess) between the eyes, and the clypeus is deflected, transverse and widely emarginate; the labrum is visible, free and lobate. The thickened side margins of the narrowed portion of the frons overarch the fronto-clypeal hollows immediately in front of the eyes. The mandibles are short and sickle-shaped; the maxillary palps are brownish yellow and the distal segment is cleaver-shaped. The eyes are large and prominent, round and convex, glassy and of a blackish brown colour. The antennae are chestnut brown and pectinate, the last nine segments being laterally elongated; the lateral elongation of the third segment is very short, the second segment is cup-shaped and the first or basal segment is very long and swollen, club-shaped. The antennae are inserted in the fronto-clypeal hollows under the thickened side margins of the frons and at the front of the eyes.

The *prothorax*, convex above and partly convex beneath, is very oblong, and its sides are largely subparallel and very narrowly rimmed or carinate; it is broadest at the base where the blunt posterior angles project considerably over the shoulders of the elytra and it is gradually narrowed at the front, being gently rounded in at the sharp anterior angles: the narrowed front is bisinuate and its breadth is about two-thirds that of the base. The *pronotum* is brownish black and thinly coated with fairly long pale brown recumbent hairs; its length (6 mm.) is slightly more than the width across the base ($5\frac{1}{2}$ mm.) and less than half the length of the elytra which is 14 mm. and it is not quite campanulate in outline, as the produced posterior angles do not project obliquely outwards; its breadth across the base is about one-third more than that of the front. The base is incurved and very declivous between the projecting angles, it has a median projecting piece which is notched for the reception of the front part of the ovate scutellum, and the posterior angles are strongly produced as two blunt-pointed projections, with

Elater fascicularis Fab. × 6

the blunt tips overlying the outer parts of the shoulders of the elytra. The sides of the pronotum are almost straight from the base to the middle and very gradually narrowed and rounded in at the sharp anterior angles, which project a little forwards at the middle of the eyes; and between the angles the narrowed front is slightly bisinuate and also a little excavate about the middle. The pronotum is finely marginate on the narrowly rimmed or carinate sides, but the edges of the front and the base appear to be plain. On the triangular area within the produced posterior angles there is a short median longitudinal ridge which extends from the irregularly ridged inner side of the angle to the disc. On the surface of the pronotal disc there are four faintly impressed foveae arranged in a square at equal distances apart, two being in a line and near the posterior angles, the other two towards the shoulders; and where the disc becomes declivous to the base there is a conspicuous median knob-like projection. Otherwise the convex discal surface is centrally smooth and shining and irregularly punctulate, the punctures larger and closer together with rugulose effect about the sides. The declivity between the median discal knob and the notched projection of the base is considerable and steep.

Beneath the pronotum the very wide and somewhat concave *prothoracic episterna* are clearly marked off by two angular and thickened sutures; the front part of each suture extends obliquely and very straight from the anterior angle to the anterior coxal cavities, and it defines the lateral boundary of the median and narrow prosternum at each side.

The *prosternum* is obconical in outline with a lip-like basal or gular lobe distinctly marked off; it is also moderately convex and it is produced between the anterior coxae into a flattened, long, narrow and tapering prosternal process which is received in the central cavity of the mesosternum. The surface of the prosternum is irregularly and not closely punctulate and is sparsely covered with hairs. The narrow middle portion of the *mesosternum* is regularly oblong, the cavity for the reception of the prosternal process being enclosed laterally and posteriorly by a strongly developed rectilinear ridge. The *metasternum* is convex, but anteriorly it is hollowed at each side of the raised median piece which meets the mesosternum in a straight line between the middle coxae. A deep median longitudinal sulcus extends between the front and hind borders of the metasternum, and the metasternal surface is finely punctulate and thinly hairy. The *metathoracic episterna* are oblong and narrow.

The *scutellum* is ovate, chestnut brown, punctulate and thinly covered with light brown hairs; it is lodged in the deep sutural recess of the elytral base.

The *elytra* are about two and a quarter times the length of the pronotum and are broadest across the middle; the rounded (arcuate) and convex bases are excavated about the suture, forming, along with the steeply sloping base of the pronotum, a large and deep recess for the scutellum; the rounded shoulder angles are a little sinuate where the pronotal posterior-angle projections rest on the elytra; the sides are slightly dilated about the middle and gradually narrowed towards the rounded and divaricate apices. The elytra are abruptly declivous in front at the base, moderately convex across the shoulders, less so behind them, rather flattened towards the apices, and finely marginate all round. The side-view outline shows gradual elevation of the surface from the apex to the short basal declivity; but the characteristic elaterid hump at the shoulders is not very prominent.

The epipleura are horizontal, broad at the base and triangularly narrowed, ending at a point on the side margins of the elytra opposite the antecoxal parts of the metasternum.

The elytra are striate-punctate, with the costa-like interstices between the rows of punctures finely punctulate. The ten rows of punctures on each elytron are more or less parallel, not regularly equidistant, and not extending to the apex. The punctures of the first or sutural row are diminished in size towards the posterior end where it joins the second row. The second and the fifth rows join anteriorly in a loop, within which is a similar loop formed by the third and fourth rows. Between and around these two loops the surface is pale orange coloured. The seventh and eighth rows unite posteriorly in a sharp loop which is closely approached by the ninth. The tenth or outermost row is very short, beginning close to the margin, at the end of the epipleuron, and joining the ninth row towards the end. The general coloration of the elytra is pale yellowish brown; portions of the interstices are pale orange coloured, the most conspicuous being the orange-coloured loop upon the base of each elytron, within which loop there is a short longitudinal band of the same colour. About the middle there is a group of dark brown longitudinal bands or fasciae of irregular size, which form a fascicular pattern about the suture and there enclose a central diamond-shaped area; behind these and towards the apex are a few similar but very short markings in the form of a continuous wavy band across the elytra and the apices are bordered with dark brown. The elytra are thinly coated with moderately long and fine pale brown recumbent hairs.

The anterior or front *coxal cavities* are entirely prosternal and open behind. The anterior coxae are round, the middle coxae are oblongish. The posterior or hind coxae, finely punctulate and lightly haired, obliquely placed and curved anteriorly, are laminate with the plates a little expanded at the inner ends; the ends are rounded and notched on the sinuate posterior borders which partly cover the large ovate trochanters and the proximal portions of the femora. All the trochanters are about the same length, but those of the hind legs are broader than the others. The short *legs* are light reddish brown and covered with pale brown hairs; the femora are moderately stout; the tarsi are five-segmented and cylindrical, the first tarsal segment is twice the length of the second, the third is shorter than the second, the fourth is the shortest, and the fifth is as long as the first. The claws are simple.

The *abdomen* is dull glossy dark chestnut brown, irregularly and not closely punctulate, and thinly coated with pale brown hairs. There are five visible sterna; the first four are about equal in length, the fifth is one fourth longer than the preceding segment and it is arcuate.

Length 19 mm.; breadth (across the middle of the elytra) 6 mm.
Hab. America (Fab. and Oliv.).
See Plate 46.

Sub-family CHALCOLEPIDIINAE

81. *Chalcolepidius porcatus* (L.) var. *virens* (Fab.)

Coleopterorum Catalogus, pars 80 (S. Schenkling, 1925), Elateridae, I, p. 58.

Mexico, Panama, Granada, Brazil.

SYN. *Elater virens* Fab., *Mant. Ins.* I, p. 172, No. 17 (1787); *Ent. Syst.* I, 2, p. 220, No. 21 (1792); *Syst. Eleuth.* II, p. 226, No. 29 (1801); Oliv. *Ent.* II, 31, p. 15, pl. 2, fig. 19 and pl. 5, fig. 55 (1790).

Fabricius, in his *Entomologiae Systematicae* refers to Voet, *Coleopt.* I, tab. 42, fig. *a*.

Two specimens in Dr Hunter's Cabinet B, drawer 7, under label

'*El. virens*
Fabr. MSS'

are co-types of this species; they answer to the very brief descriptions of *virens* given by Fabricius and Olivier. Olivier's fig. 55, although roughly drawn, obviously corresponds with the larger co-type; and his fig. 19 agrees with the smaller one, except that the coloration is represented as uniform light blue. The larger co-type is a female, now rather dusty but scarcely rubbed, and it agrees well with a female from Grenada, West Indies, in the British Museum Collection.

A series of modern examples in the Bishop Collection, which agree more or less closely with the co-types, show a considerable range in colouring, from bright green with creamy white to dark green with bluish tinges; one of the specimens has blue elytra.

Description of Co-type, *Elater virens* Fab. This, the larger co-type, is a *female*. Form elongate-oblong, of a long oval shape, ellipsoidal transversely, the outline of the regions regularly continuous; moderately convex above towards the head and over the elytra, but depressed on the head and about the scutellar area and flattened along the rim-like sides of the thorax. The elytra slightly and regularly costate and striated, with rows of spaced-out punctures in the interstices and with very fine scattered punctules on the costae. The eyes round, convex and glassy green, not very prominent; the antennae dull black and serrate. The entire surface, above and below, also the legs and the antennae, glossy greenish black, but almost completely coated with very short and thick hairs, so closely overlying as to produce a finely shagreened appearance mainly golden green and partly blue. The bare median portion of the undersurface finely punctulate.

The *head*, insunk in the thorax close up to the eyes, is nearly square, but it is rounded off and narrowed in front. The vertex is broad and a little depressed, the narrowed frons is hollowed (frontal recess) between the eyes; the clypeus is deflected and transverse, and the labrum is visible, transverse and fringed with bristle-like hairs. The thickened sides of the frons and clypeus form outjutting arches over the fronto-clypeal hollows immediately in front of the eyes. The mandibles are short and angulate, the maxillary palps are dark green and the distal segment is cleaver-shaped.

The eyes are rounded and rather convex but not conspicuously large, not prominent, and they are glassy green.

The serrate antennae are inserted in front of and near the eyes in the fronto-clypeal hollows and under the thickened fronto-clypeal margins. The basal or first antennal segment is the longest and is swollen (club-shaped), the second segment is very small and rounded, the succeeding eight segments are serrate and the terminal or eleventh segment is lobate. The antennae are dull greenish black, except the first three segments, which are shagreened golden green. The serrate angle of the antennae is distinctly acute, less than a right angle.

The *prothorax* is very oblong, it is broadest at the base, where the posterior angles project a little outwards; towards the front it is gently narrowed and there rounded in at the anterior angles; it is deeply and widely excavate, and also slightly bisinuate between the anterior angles, which project forwards; its width in front is about half that of the base; it is convex above and concave beneath at each side of the convex prosternum, and its sides are expanded in narrow and thick-edged rims.

The *pronotum* is campanulate in outline, its length is about half that of the elytra and slightly greater than its width across the base. The breadth across the expanded base is about twice that of the front; the base is incurved and has a median projecting piece with a deep rounded notch for the front part of the scutellum; the posterior angles, which project a little outwards, are rounded at the sides and inturned so as to be regularly continuous with the elytra at the shoulder angles. The sides of the pronotum are gradually narrowed and rounded in at the blunt anterior angles, which are produced alongside the lower portion of the eyes; and between the anterior angles the narrowed front is widely excavate and distinctly bisinuate. The pronotum is marginate on the sides and around the four angles; the thick-margined sides are carinate, thus forming somewhat expanded rims. The triangular areas within the posterior angles are centrally a little elevated and therefore not quite flat, and from there the disc is gently raised, gradually becoming more convex towards the front where it is raised over the head. The surface of the pronotal disc is smooth, save for two transverse linear border impressions in front (opposite the eyes) and a slight but distinct median longitudinal ridge; this ridge extends over

Elater virens Fab. $\times 3\frac{1}{2}$

the disc from the front to the base where it is forked and divergent to each side of the median notch. The glossy black surface, which is closely pitted with shallow pits and also irregularly punctate, is thickly coated with very short thick hairs and has a golden green shagreened appearance.

Beneath the pronotum the wide and concave *prothoracic episterna* are clearly defined by the two waved sutures which mark the lateral boundary of the median prosternum at each side and which extend obliquely inwards from the deep prosternal grooves (for the antennae) to the anterior coxal cavities. The coloration of the proepisterna is golden green and blue.

The convex *prosternum*, broad and thick-rimmed at the base, is gradually narrowed to the anterior coxae, and it is produced between them into a tapering blunt-pointed and glossy black prosternal process which fits into the central cavity of the small and hollowed *mesosternum*.

Anteriorly the long and broad *metasternum* is produced between and in front of the middle coxae into a two-prong process based on the mesosternum, the long notch between the prongs being vertically continuous with the central mesosternal cavity. Posteriorly, the oblique base of the metasternum has a median semicircular lobe with a small round notch-like depression in the middle of its edge. The metasternum is convex with the sides hollowed anteriorly, and it is marked by a median longitudinal depressed line which begins near the basal lobe and extends to the notch in front; the convex surface about this line is flattened, smooth and glossy black. The *metathoracic episterna* are oblong and narrow.

The *scutellum* is obcordate, narrow and rounded, and emarginate in front; it is broad and rounded and very slightly emarginate behind. Its surface is flush with that of the elytra, the depression in which it is lodged not being deep.

The *elytra* are about twice the length of the pronotum and are broadest about the middle; the width across the base is the same as that of the pronotum; the bases are sharply angulate near the scutellum, and the shoulder angles are obtuse angles; the sides, a little constricted just behind the shoulder angles, are slightly rounded out about the middle and gradually narrowed towards the apices, which are rounded in and horizontal (a little sinuous) with the sutural tips divergent and slightly projecting.

The elytra are moderately convex and markedly raised with gibbous effect near the base, the surface being gradually elevated from the apex to the short basal declivity. The convex surface is a little depressed along the suture. From about the middle of the sides to the suture the elytra are narrowly rimmed; and they are finely marginate all round except at the epipleura where the margin is interrupted, a part of it being obsolete. The shoulders are not prominent, not very callose. The elytral surface is slightly and regularly costate, and between the costae it is striate with single rows of punctures which are at intervals apart; the costal surfaces are very finely punctulate. The costae and striae extend to the apex, excepting the three outermost striae, which unite terminally before the apex. The epipleura are oblong; at the posterior coxae they are curved in and gradually narrowed, vanishing at a point on the elytral side margins opposite the middle of the second abdominal sternum. The coloration of the elytra is golden green with blue about the margin and along the outer sides; the epipleura are blue.

The anterior or front coxal cavities are entirely prosternal and open behind. The *legs*, closely covered with very short hairs, are golden green with a bluish tinge. The anterior coxae are round, the middle coxae are roughly oblong. The posterior or hind coxae are obliquely placed and laminate, with the broad ends rounded and notched on the sinuate posterior borders, which partly cover the large trochanters and the proximal portions of the femora. All the trochanters are large and about the same length. The femora are stout, particularly the front and middle femora, which are arcuate;

the hind femora are straight. The front and middle tibiae are furnished with long hairs beneath. The tarsi are five-segmented and the first four tarsal segments are laterally lobed and flat beneath; the first segment is twice the length of the second, the third is shorter than the second, and the fourth is the shortest; the fifth segment is club-shaped and it is the longest. The claws are simple.

The *abdomen* is glossy black and finely punctulate along the middle, the lateral portions are golden green and coated with short close hairs. There are five visible sterna and the first four are of equal length; the fifth is one and a half times the length of the preceding sternum, it is arcuate and its rounded posterior border is thickly fringed with black bristly hairs.

Length 40 mm.; maximum breadth (across the middle of the elytra) 15 mm.

Hab. In American Islands (Fab.), South America, in Brazil at Carthagena (Oliv.).

See Plate 47.

The smaller co-type is uniformly golden green above and beneath, greenish black on the bare parts; its length is 33 mm.

Super-family *Heteromera*

Family Tenebrionidae

The following are the species of Tenebrionidae mentioned by Fabricius, in his published works, as having been described by him from specimens in Dr Hunter's Collection:

Erotylus nebulosus *Sp. Ins.* I, p. 158, No. 10 (1781).
Pimelia gibba *Mant. Ins.* I, p. 207, No. 3 (1787).
Helops longipes *Ibid.* p. 214, No. 14 (1787).
Cnodulon nebulosum *Syst. Eleuth.* II, p. 13, No. 2 (1801).

The above names are the original names as given by Fabricius, and the references are to the works in which these species were first described.

The following species were described by Olivier from examples in Dr Hunter's Collection:

Diaperis bicornis *Ent.* III, 55, p. 6, pl. 1, figs. 4*a* and *b* (1795).

Tenebrio serratus *Ibid.* 57, p. 5, pl. 1, fig. 1 (1795).

82. *Pyanisia undata* (Fab.)

Coleopterorum Catalogus, pars 28 (H. Gebien, 1911), Tenebrionidae, III, p. 585.

Tropical America.

SYN. *Erotylus nebulosus* Fab., *Sp. Ins.* I, p. 158, No. 10 (1781); *Mant. Ins.* I, p. 92, No. 16 (1787); *Ent. Syst.* I, 2, p. 40, No. 22 (1792).
Cnodulon nebulosum Fab., *Syst. Eleuth.* II, p. 13, No. 2 (1801).
Helops nebulosus (Fab.), Oliv. *Ent.* III, 58, p. 11, pl. 2, fig. 3 (1795).
H. undatus Fab., *Ent. Syst.* I, 1, p. 122, No. 25 (1792); Oliv. *Ent.* III, 58, p. 11, pl. 2, fig. 4 (1795).

The single example in Cabinet A, drawer 4, under label

'*Erot. nebulosus*
Fabr. pag. 158, No. 10'

is evidently the type; it has been compared with a British Museum modern example and also with one from the Bishop Collection, and they correspond closely. This type answers the descriptions of *nebulosus* and *undatus* given by Fabricius and Olivier, but neither of Olivier's figures exactly represent it; the markings on its thorax are indicated in his figure of *undatus* and those on the elytra are shown in his figure of *nebulosus*. According to Dr K. G. Blair, Olivier's figures obviously portray lightly and more heavily marked individuals of the one species. Gebien places this species under *Epitragus* (*Coleopterorum Catalogus*, pars 15, Tenebrionidae, I, p. 25), but Blair considers this an error.

Description of Type, *Erotylus nebulosus* Fab. Form oblong-ovate, almost continuous in outline, the sides largely parallel, and convex above; the head flattened and somewhat overarched by the convex pronotum. The head black, except the narrow band-like clypeus, which is bright yellow green; the antennae black, and the first five antennal segments nitid (shining). The thorax smooth, slightly bordered with black and with three irregular oblong black patches side by side on the disc and surrounded with red. The elytra black with three dull brick-red wavy transverse bands and with two red apical spots oblong and oblique; the first band branched (each side) around two small black areas. The elytral surface lightly punctate-striate, the four middle rows of punctures interrupted at the posterior edge of the first red band and there forming two loops; the rows recontinued from loops around the two black areas within the branches of the red band. The under surface of the body metallic black and smooth with the terminal sterna coppery. The long legs black, smooth and somewhat coppery.

The *head*, dull glossy black and irregularly punctate, is broadly set on the thorax and somewhat insunk; it is dilated at the sides, narrowed in front and flattened above. The vertex is narrowed between the eyes, which encroach on it, and is marked by a slight median ridge; the frons is elevated at the sides, forming strong lateral outjutting arches over the antennal sockets, and it is divided by a transverse semicircular line across the recess between the lateral ridges; the clypeus is a conspicuous yellow and narrow transverse band between the bases of the mandibles; the labrum is a transverse lobe fringed with golden hairs; the maxillary palps are exserted and clavate, the last segment is dilated and apparently obconical. The large transverse gula is dull black, closely punctate and rugulose, except a small median posterior lobe-like part which is yellowish brown and smooth and clearly marked off by two short deeply impressed lines (gular sutures) which converge but do not meet above the lobe.

The conspicuous bright pale yellow eyes, which have large

hexagonal convex facets, are transverse and emarginate (reniform), extending from the narrow genae and encroaching on the vertex. The eleven-segmented antennae, filiform and gradually thickened towards their tips, are inserted within the emarginations of the eyes, in the frontal hollows under the thickened overarching lateral margins of the frons. The first or basal segment of the antennae is clavate; the second, which is the smallest of the series, is very small and moniliform; the third, fourth and fifth are clavate; segments six to ten are subconical, and the terminal one is oval. The first five antennal segments are glossy black, and the last six are dull black and strongly pitted.

The *pronotum* is transverse; it is broadest at the base, which is as wide as the base of the elytra, and the sides are gently rounded in towards the narrowed front, which is a little concave. The base is sinuate, with a short but wide median scutellar lobe; the posterior angles are rounded and the anterior angles are sharp. In front and along the sides the pronotum is marginate; its surface is markedly convex and smooth with a dull gloss; it is completely bordered with black, and within this border, which is narrow and irregular, the disc is dull brick-red around three large irregular oblong black patches side by side; the central and largest patch has narrow connections with the two lesser patches and with the front portion of the marginal border. Between the middle and the base of the disc, and just within the central black patch, there are two small pits, like large punctures, parallel but wide apart.

The *prothoracic episterna*, which extend to the anterior coxal cavities, are large and triangular, partly convex and dull black. The large dagger-shaped prosternal process of the narrow transverse *prosternum* is partly lodged in the median triangular recess of the small and hollowed *mesosternum*. The moderately long, broad and convex *metasternum* is dull black and smooth; it is marked by a median longitudinal impressed line which passes into the sulcus at the notch in the middle of

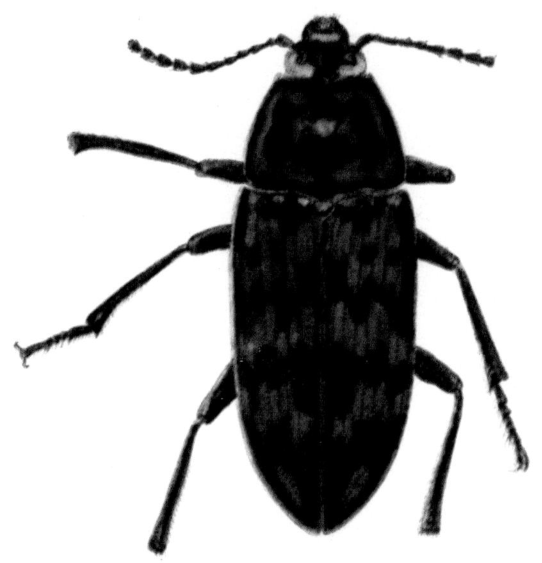

Erotylus nebulosus Fab. × 6

the posterior border of the metasternum. The *metathoracic episterna* are oblong and narrow, and a distinct oblique line marks off the small *metepimera.*

The *scutellum* is a little insunk; it is broad and shield-shaped, and its surface is flattened and smooth.

The *elytra* are entire, convex, markedly arcuate in outline posteriorly (lateral view), and finely marginate along the outer sides. The base is bisinuate; and behind the prominent shoulder angles, which form short projections, the sides are a little rounded out, thence straight and parallel about the middle, and gradually incurved on the apical third towards the convergent apices. The elytral surface is regular, except at the base where the surface is depressed around the scutellum and also slightly between the scutellar elevation and the prominent shoulder callus of each elytron. The elytra are black with three dull brick-red transverse bands and with a red oblong apical spot on each elytron; the bands are irregular and sharply sinuous, and the first one is branched where it surrounds two small black spot-like areas. There are nine subparallel rows of lightly impressed punctures and a short sutural stria of stronger punctures on each elytron; the punctures are within pale-coloured spots. The outermost row of punctures is along the elytral margin; and the rows unite apically, forming sharp loops within each other. The four middle rows are interrupted at the posterior edge of the first or proximal red band, there forming two loops; these rows are recontinued to the base from loops around the two small black areas within the branched red band. The *epipleura*, dull black and smooth, are wide, narrowing towards the elytral apices.

The anterior coxal cavities are entirely prosternal. The *legs* are long, black above and coppery beneath, and punctulate. The anterior or front coxae are globose, the middle coxae are rounded, the hind or posterior coxae are transverse and not widely separated; the antecoxal piece is rib-like. The trochanters of the front legs are larger than those of the middle and hind legs, which are of equal size. All the femora are of equal length and moderately stout; the tibiae are slender, gradually widened towards the distal end, furnished with short hairs beneath and with two short distal spurs. The front and hind tarsi are wanting in this type specimen; the middle tarsi are five-segmented, the first or proximal segment is about equal in length to the second, third and fourth together, which are small and equal-sized; the fifth or distal segment is longer than the first, it is club-shaped and it bears two simple claws. The under surfaces of the tarsi have a golden pubescence.

The *abdomen* is dull glossy black, except the terminal segments, which are bright coppery; the surface of the segments is finely punctulate. There are five abdominal sterna; the first and second sterna are about equal in length; the third and fourth sterna are shorter, each being about half the length of the second; the fifth is nearly as long as the second, its posterior border is arcuate and is fringed with golden hairs.

Length 15 mm.; breadth (across middle of elytra) 6 mm.
Hab. In South American Islands, Cayenne (Fab. and Oliv.).
See Plate 48.

83. *Pimelia gibba* (Fab.)

Coleopterorum Catalogus, pars 22 (H. Gebien, 1910), Tenebrionidae, II, p. 208.

Algeria.

SYN. *Tenebrio gibbus* Pallas, *Icon.* I, p. 46, tab. c, fig. 12 (1781).
Pimelia planata Thunb., *Nov. Ins. Sp.* p. 6, fig. 120 (1784).
P. gibba Fab., *Mant. Ins.* I, p. 207, No. 3 (1787); *Ent. Syst.* I, I, p. 100, No. 3 (1792); *Syst. Eleuth.* I, p. 128, No. 4 (1801); Oliv. *Ent.* III, 59, p. 6, pl. 2, fig. 24 (1795).
P. simplex Sol., *Ann. Soc. Ent. Fr.* V, p. 123 (1836).
Melanostola gibba (Fab.), *Geb. Deutsche Ent. Zeitschr.* p. 228 (1906).

The specimen in Cabinet B, drawer 12, under label

'*Pim. gibba*
Fabr. MSS'

is the type; it closely agrees with a modern example in the British Museum Collection, with which it has been compared. Dr K. G. Blair made the following note with reference to this insect: 'The specimen in the Kiel Museum identified by Gebien (1906) with this species and with *Pimelia simplex* Sol. is not the type and cannot agree with the description "elytris linea elevata abbreviata".'

Description of Type, *Pimelia gibba* (Fab.). Form robust, anteriorly globular and posteriorly obovate, constricted between. Uniformly dull glossy black, except the lower parts of the legs, which appear to be pitchy brown. The head vertical and regularly continuous with the high front of the globular thorax which conceals it from upper view; the abdomen obovate and almost entirely covered by the high gibbous elytra, the sides and the apical portions of which are steeply declivous, like the head and the front of the thorax. Upon each elytron there is a strongly raised and corrugated longitudinal ridge which becomes obsolete at the declivity.

The *head* is relatively small and vertical; viewed from the front it is triangular, and it is insunk in the thorax up to the eyes; the frons is a little convex and irregularly punctulate; the short and narrowed clypeus is slightly arcuate in front, it is marked off from the frons by a faint linear depression between the raised and outjutting lateral fronto-clypeal rims, and it is punctulate. The fronto-clypeal rims form an angular junction with the thickened sinuate rim of the gena, thus overarching and enclosing the angular antennal sockets, which are just behind the broadly expanded bases of the mandibles. The short transverse labrum is narrower than the clypeus; it is overarched by the clypeus, and it has a thick fringe of golden hairs. The broad-based thick mandibles are apparently bifid or bidenticulate. The finely faceted eyes are wide apart, small and inconspicuous, kidney-shaped; and the thickened angle of the antennal socket overlies the eye notch or emargination.

The *prothorax* is globular, each way (i.e. length and width) it is about half the size of the abdomen. The *pronotum* is cap-like, and the front portion is high and vertically declivous. The pronotum is distinctly bordered all round; the narrow front margin is incurved, the narrow margins of the sides are arcuate and slope downwards to the sides of the head, the base is bordered by a thickening which is slightly arcuate and confluent at the rounded posterior angles with the narrow side margins. The anterior angles have the form of short blunt-pointed projections. The pronotal surface is punctulate about the middle and front of the disc, and is rugulose about the sides and at the base, strongly so around the side margins. The wide and convex *prothoracic episterna* are marked off by the faintly impressed sinuous sutures which define the lateral boundary of the median prosternum and the anterior coxal cavities at each side. The surface of the proepisterna is slightly rugulose.

The *prosternum* is very convex and roughly T-shaped; its base is short and broad, with an incurved and thickened rim, and it is produced between the anterior coxae into a large oblong prosternal process which is very convex, its posterior portion being so turned in that it faces against the inflected part of the mesosternum. The process is thick-rimmed, and its surface is marked with short and irregular linear impressions. The *mesosternum* is convex and short, and is inflected in front; the sides are oblique with thick rims and the surface is a little rugose. A slightly elevated transverse line in the hollow between the middle coxae marks off the mesosternum from the metasternum. The central portion of the *metasternum* is a small convex disc; the middle portions between the coxal cavities each side are thick-rimmed and very narrow, consequently the middle and hind coxae are close together; the lateral portions are expanded, triangular and moderately convex. The surface of the metasternum is lined on the central disc and punctulate on the lateral parts.

Pimelia gibba (Fab.) × 4

This specimen is badly worn and damaged about the bases of the elytra, and there is now no trace of the *scutellum*.

The *elytra* are entire, obovate (viewed from above), and lofty or gibbous, being very strongly convex with steep sides which are almost perpendicular. Viewed from the side, the elytra are arcuate in outline with the apical portions vertically declivous. The narrowed base of the elytra is the same width as the base of the pronotum; the shoulder angles are rounded and the sides are marginate, having narrow projecting rims. Viewed laterally, these rims are arcuate and so the *epipleura* are semicircular, broad about the middle and narrowed at each end. The sutural edges of the elytra are thickened and a little corrugated, except towards the apices. On each elytron a strongly raised and corrugated ridge extends from the shoulder to the apical declivity, there becoming obsolete; this conspicuous ridge is in great part parallel with the suture, and it forms the upper boundary of the steep side.

The elytral surface is slightly rugulose here and there; on the apical area it is irregularly studded with small tubercles, which are not very close together and which gradually diminish in number and size and disappear towards the middle of the back.

The anterior coxal cavities are entirely prosternal, closed behind; the anterior or front coxae are globose and somewhat prominent; the intermediate or middle coxae are rounded and a little transverse; the hind or posterior coxae are ovate, transverse and farther apart than the middle coxae.

The *legs* are long. All the trochanters are about the same size. The femora are moderately stout; the front femora are the stoutest and shortest and the hind femora are the longest, being one-third longer than the middle femora and nearly twice the length of the front femora, and they are arcuate. The femoral surface is rugulose and punctate with short golden hairs. The tibiae equal the respective femora in length, but they are much thinner, and each has two distal spurs. The tibial surface, closely punctate, is set with short sharp spines and golden hairs. The tarsi are shorter than the tibiae; the front and middle tarsi are five-segmented. The first four tarsal segments are conical; the first or basal segment is about as long as the second, third and fourth together, which are short and of nearly equal size; the fifth is a little longer than the first and is club-shaped, and it bears two simple claws. The hind tarsi are wanting.

The *abdomen* has five visible sterna; the first, second and fifth are about equal in length and are the longest; the third is shorter than the second, and the fourth, which is the smallest, is about half the size of the second. The surface of the first abdominal sternum is alutaceous, that of the second less so, and the minute crack-like marks are mostly longitudinal; the surface of the third, fourth and fifth sterna is punctulate; and the posterior border of the fifth is regularly arcuate.

Length 19 mm.; breadth (across the middle of the elytra) 11 mm.
Hab. India (Fab.), India and Africa (Oliv.).
See Plate 49.

84. *Hoplocephala bicornis* (Fab.)

Coleopterorum Catalogus, pars 28 (H. Gebien, 1911), Tenebrionidae, III, p. 367.

North and Central America.

SYN. *Hispa bicornis* Fab., *Gen. Ins.* p. 215, Nos. 3–4 (1776); *Sp. Ins.* I, p. 82, No. 4 (1781); *Mant. Ins.* I, p. 47, No. 4 (1787).
Diaperis bicornis Oliv., *Ent.* III, 55, p. 6, pl. 1, figs. 4*a* and *b* (1795).
Hoplocephala metallica Beauv., *Ins. Afr. et Am.* p. 139, t. 30*b*, f. 2 (1805).
H. virescens Cast. et Brll., *Ann. Sci. Nat.* XXIII, 1831, p. 341 (17) (Mon.).
H. gracilis Motsch., *Bull. Mosc.* XLVI, p. 467 (1873).

The specimen in Cabinet A, drawer 3, under label

'*Hisp. bicornis*
Fabr. p. 82, No. 4'

is presumably the insect described and figured by Olivier as *Diaperis bicornis*, it closely agrees with his description and figures and also answers the description given by Fabricius; and it corresponds with examples of this species in the British Museum Collection.

The Fabrician type is stated to be in the Yeats Collection, and it is possible that the Hunterian example is the Yeats insect.

Description of Type, *Hispa bicornis* Fab. Form oblong-ovate, the outline of the thorax and abdomen almost regularly continuous, moderately convex above; the head relatively small and flattened with two conspicuous upright simple horns and with short antennae, moniliform in general appearance but perfoliate. Glossy bronze green above, reddish brown beneath, the antennae pitchy brown and rusty red, the legs ferruginous. Pronotum punctulate, the elytra slightly costate with rows of punctures (punctate-striate) between the costae; the underparts punctate.

The *head* is bronze green, short and transverse and broadly set on the thorax which overarches the vertex close up to the eyes; it is flattened and a little insunk on the frons, which is narrowed between the eyes and arcuate in front; the eye margins of the frons are raised and each bears a short erect simple and pointed horn, a little longer than the head, and on the middle of the arcuate front margin there are two small pointed tubercles a short distance apart; the clypeus is a narrow transverse strip below the frons and between the bases of the mandibles; the labrum is transverse, semicircular, ferruginous and fringed with golden hairs; the maxillary palps are exserted and clavate, the last segment is enlarged and obconical. The gula is transverse and convex, reddish brown and punctate, except a small median posterior part which is triangular and orange-red. The eyes are not conspicuous, relatively small, reniform and coarsely faceted. The short eleven-segmented antennae, which have a somewhat moniliform appearance, are perfoliate and are ferruginous about the base and brown towards the tip; the first three basal segments are narrow, the succeeding segments are broader, cupuliform and about equal in size, the terminal segment is larger and conical. The antennae are inserted immediately in front of the eyes and under the outjutting ends of the thickened arcuate front margin of the frons.

The *pronotum* is transverse, about twice as broad as long, and moderately convex; in front it is much broader than the head, it is broadest at the base, and it is finely marginate all round. The front margin of the pronotum is slightly and widely excavate and distinctly sinuate; the rounded side margins are a little angulate behind the middle; the hind margin is sinuate with a wide median scutellar lobe, the anterior and posterior angles are rounded, and the surface of the disc is irregularly punctulate. The broad, triangular and convex *prothoracic episterna* are clearly defined by sinuous sutures which form the lateral boundary of the prosternum and the anterior coxal cavities at each side; the surface of the proepisterna is closely punctate with some confluence.

The convex *prosternum* is comparatively small; its basal portion is transverse and crescentic, with a thickened front rim continuous on the proepisterna; this sterno-episternal rim forms the deeply concave margin behind the gula. The prosternal process between the anterior coxae is short and narrow. The *mesosternum* is convex, short and punctate with golden hairs; it is marked off from the metasternum by an elevated and deeply bisinuate suture. The *metasternum* is transverse and moderately convex, straight at the sides and sinuate behind, and clearly defined by the elevated enclosing suture. The posterior part of the metasternum is marked by a broad longitudinal sulcus with a median impressed line, and the surface is punctate with short and fine golden hairs. The *metathoracic episterna* are distinctly marked off as narrow oblong punctate plates; the small *metepimera* are oblique and smooth.

The *scutellum* is broad and shield-shaped, with a slightly

Hispa bicornis Fab. × 30

raised disc, and apart from a few punctures the surface is smooth.

The *elytra* are entire, moderately convex, somewhat flattened about the suture anteriorly. The base of the elytra is about the same breadth as the base of the pronotum and is sinuate; the shoulder angles are rounded; the sides, which are in great part subparallel, are gently rounded in and narrowed towards the blunt-pointed apices. The shoulder callosities are distinct, but not very prominent. The elytra are finely marginate, except along the inner or sutural edges. The elytral surface is slightly costate and punctate-striate; on each elytron there are nine subparallel rows of punctures and a short oblique scutellar row, and between the rows, which extend from the base to the apex, there are some scattered punctules. The *epipleura* are relatively wide, and the width is regular except near the elytral apices.

The anterior coxal cavities are entirely prosternal, closed behind. The *legs* are ferruginous and rather short. The anterior or front coxae are globose and fairly prominent; the middle coxae are rounded and transverse; the posterior or hind coxae are transverse and farther apart than the middle coxae. The femora are stouter than the tibiae, which are nearly as long as the femora, and each tibia bears two very short distal spines. The tarsus of the front leg is shorter than the tibia and five-segmented; the first four tarsal segments are roughly conical; the first or basal segment is as long as the second, third and fourth together, which are short and of almost equal size; the fifth is a little longer than the first and is club-shaped, and it bears two simple claws. The tarsi of the middle and hind legs are wanting.

The *abdomen* has five visible sterna; the first and second, which are of equal length, are the longest; the third is shorter than the second, the fourth is about half the length of the third; the fifth, which is as long as the third, is marked anteriorly by a transverse furrow and its posterior margin is regularly arcuate. The surface of the abdominal sterna is punctate, and there is also a faint median longitudinal depression.

Length 3½ mm.; breadth 1½ mm.
Hab. North America (Fab. and Oliv.).
See Plate 50.

85. *Prioscelis serrata* (Fab.)

Coleopterorum Catalogus, pars 28 (H. Gebien, 1911), Tenebrionidae, III, p. 475.

West Africa.

SYN. *Tenebrio serratus* Fab., *Syst. Ent.* p. 255, No. 1 (1775); *Sp. Ins.* I, p. 322, No. 1 (1781); *Mant. Ins.* I, p. 211, No. 3 (1787); *Ent. Syst.* I, I, p. 111, No. 4 (1792); *Syst. Eleuth.* I, p. 145, No. 6 (1801); Oliv. *Ent.* III, 57, p. 5, pl. 1, fig. 1 (1795).

The Fabrician type of this species is in the Banks Collection,[1] British Museum. The Hunterian specimen in Cabinet B, drawer 13, under label

'*Ten. serratus*
Fabr. pag. 322, No. 1'

has been compared with the type and it closely corresponds; it is the insect figured and described by Olivier. In his description, Olivier does not mention that the elytral striae are punctate.

86. *Eupezus longipes* (Fab.)

Coleopterorum Catalogus, pars 28 (H. Gebien, 1911), Tenebrionidae, III, p. 574.

Guinea.

SYN. *Helops longipes* Fab., *Sp. Ins.* I, p. 326, No. 10 (1781); *Mant. Ins.* I, p. 214, No. 14 (1787); *Ent. Syst.* I, I, p. 121, No. 20 (1792); *Syst. Eleuth.* I, p. 161, No. 34 (1801); Oliv. *Ent.* III, 58, p. 14, pl. 2, fig. 8 (1795).

[1] 'On the Fabrician Types of Tenebrionidae (Coleoptera) in the Banks Collection' by K. G. Blair (*Annals and Magazine of Natural History*, Vol. XIII, Eighth Series, May 1914).

The type of this species is in the Banks Collection (British Museum); but in his *Mantissa Insectorum* Fabricius refers to another example: 'Alium simillimum vidi in Musaeo D. Hunter at tibiis omnibus laevibus', which therefore may be regarded as a metatype. There are two specimens in Cabinet B, drawer 13, under label

'*Hel. longipes*
Fabr. pag. 326, No. 10'

One of the specimens is a male and the other, the Fabrician metatype, is a female; they answer the descriptions given by Fabricius and Olivier. The male, which is remarkably perfect, obviously is the insect figured and described by Olivier.

Description of Metatype, *Helops longipes* Fab. The example is a *female*. Form elongate-ovate, stout and moderately convex, almost regularly continuous in outline, and gibbous in side view. The head relatively small, insunk within the thorax and vertical, scarcely visible from above, and with long filiform antennae. The prothorax transverse; the elytra entire, covering the abdomen, partly subparallel, striate-punctate and close fitting to the blunt tips; the legs very long, with rather slender tibiae. Concolorous dull black above; the head, antennae, limbs and sterna glossy dark chestnut brown.

The glossy dark brown *head*, vertical and deeply insunk in the prothorax, is broad rather than long. The vertex is hidden from view beneath the overlapping pronotum. The frons is very flattened, its surface is largely occupied by the eyes; the upper portion is small and triangular, the middle portion is represented by a very narrow strip, and the lower part (fronto-clypeus) is transverse and hollowed, with prominent lateral ridges immediately in front of the middle of the eyes and overhanging the antennal sockets. The clypeus is narrow and transverse, smooth and impunctate; the labrum is a transverse lobe as broad as the clypeus and fringed with bright golden hairs; the stout mandibles are of the herbivorous type; the maxillary palps are short and stout, with the terminal segment enlarged and triangular (securiform).

The eyes are golden green, reniform, deeply notched and close together in front; they encroach so greatly on the frons that there is only a narrow median frontal strip between them. The eleven-segmented filiform antennae are long (15 mm.), glossy dark brown, and inserted under the fronto-clypeal ridges immediately in front of the middle of the eyes. The first antennal segment, the stoutest of the series, is clavate; the second, which is moniliform, is the shortest; the third is the longest and the fourth is half the length of the third; the fifth, sixth and seventh segments are of equal size, the length of each being two-thirds that of the third; the last four segments are also equal, they are each one-half the length of the seventh, but distinctly stouter.

The *prothorax* is dull black, declivous and wedge-shaped in side view. The *pronotum* is transverse, moderately convex, unevenly rounded at the sides and narrowed to the front; it is marginate except along the base, and its surface is regular and smooth, very finely punctulate; the narrowed front is subcrescentic, being widely but not deeply excavate, and slightly sinuate; the base is not as broad as the elytra, and it is sinuate on each side of the median scutellar portion, which is straight and has the same width as the straight base of the scutellum. The anterior and posterior angles are sharp obtuse angles, and the lateral borders are sinuate. The large propleura are wedge-shaped, deep, convex, smooth and finely punctulate; the mesopleura are small and roughly rectangular, the *mesepisternum* is somewhat triangular and smaller than the transverse *mesepimeron*; the metapleura (*metepisterna*) are large, oblong, and a little hollowed, and the surface, like that of the other thoracic pleura, is sparsely and very finely punctulate.

The *prosternum* is small; but the prosternal process, extending between and beyond the anterior coxae, is large, rhomboidal and hollowed on each side of a prominent median longitudinal ridge. The *mesosternum* is small, and it is deeply grooved in front of and between the middle coxae. The *metasternum* is large, transverse, very convex and marginate in front; it has a broad anterior lobe extending between the middle coxae, and it is marked by a median longitudinal impressed line; the sternal surface is very smooth and finely punctulate, and the antecoxal pieces are distinct and narrow.

The *scutellum* is dull black, triangular and centrally depressed.

The *elytra* are dull black, broader at the base than the pronotum, entire and close fitting, completely covering the abdomen, and strongly convex; they are rounded in at the base, the sides are subparallel to the apical third and there rather sharply rounded in to the suture, and the apices are blunt; the outer borders are inflexed and marginate, and the shoulders are prominent.

The surface of the elytra is striate-punctate, the striae are clearly impressed and are disposed at fairly regular intervals apart, and the punctures are not very close together. There is a distinct scutellary striole between the suture and the first stria. The first or sutural stria forms a loop anteriorly with its neighbour; all the striae converge in the apical area and join in pairs, thus forming sharp posterior loops. The interstices between the striae are thinly and very finely punctulate. The *epipleura* are broad at the shoulders.

The *legs* are long, glossy dark brown, smooth and finely punctulate; the front pair are the shortest, the hind pair are the longest, the tibiae of the hind legs being longer than the femora. The front coxae are short and globose; the middle coxae, which are rounded but not globose, are less prominent and farther apart than the front coxae; the hind coxae are transverse and the same distance apart as the middle coxae. The femora are not very stout. The tibiae are rather slender, subcylindrical, and distinctly widened out at their distal ends which are ventrally covered with reddish hair; and they have each two very small apical spurs.

The front and middle tarsi are five-segmented; the front tarsus is two-thirds of the length of the tibia, the first four tarsal segments are short and about equal in size, the fifth segment is long, fully half the length of the tarsus; the middle tarsus is two-thirds of the length of the tibia and the first segment is long, its length is equal to that of the short second, third and fourth segments taken together, and the fifth is longer than the first but less than half the length of the tarsus; the hind tarsus is about half the length of the tibia, the first and fourth segments are equally long, and the second and third segments are short. All the tarsi are ventrally covered with reddish golden hair, and the claws are simple.

The *abdomen* is very convex and glossy dark brown; its surface is faintly rugulose and finely punctulate. There are five visible sterna; the first abdominal sternum has a large and lobate intercoxal process between the hind coxae, the second sternum is longer than the first and the third, the fourth is the shortest, and the fifth, which is as long as the second, is narrowed and lobe-like, broadly rounded at the tip.

Length 22 mm.; breadth (across the base of the elytra) 12 mm.
Hab. Central Africa (Fab. and Oliv.).

The *male* example, described and figured by Olivier, is 20 mm. long and 11 mm. broad; and the length of the antennae is 17 mm. The middle portions of the posterior borders of the third and fourth abdominal sterna are thickened and tuberculate; and the inner or under side of the front and middle tibiae (except the anterior third) is villose with raised and closely set red hair, long on the middle tibiae.

A specimen labelled N. Zululand, in the Bishop Collection, has the hind as well as the front and middle tibiae villose with red hair.

Family Oedemeridae

87. *Copidita lateralis* Ch. O. Waterh.

Trans. Ent. Soc. London, 1878, p. 307. *Coleopterorum Catalogus*, pars 65 (S. Schenkling, 1915), Oedemeridae, p. 28.

San Domingo, Jamaica.

Syn. *Lagria vittata* Fab., *Syst. Ent.* p. 125, No. 5 (1775); *Sp. Ins.* I, p. 159, No. 6 (1781); *Mant. Ins.* I, p. 93, No. 6 (1787).
Dryops vittata Fab., *Ent. Syst.* I, 2, p. 76, No. 7 (1792); *Syst. Eleuth.* II, p. 68, No. 7 (1801).
Oedemera vittata Oliv., *Ent.* III, 50, p. 7, pl. 1, fig. 6 (1795).

The specimen with attached label '*Lagria vittata*' (the handwriting presumably Olivier's) in Cabinet A, drawer 4, is probably the insect described by Olivier as *Oedemera vittata*; it corresponds with the figure he gives and answers his description, except that the scutellum is large rather than small. Dr K. G. Blair examined this type and identified it as *Copidita lateralis*. The elytral costae described by Mr Waterhouse are not distinct in this Hunterian example, and the broad grey-purple vitta or stripe on each elytron does not extend to the apical margin.

The above synonymy is that given by Olivier.

Description of Type, *Oedemera vittata* Oliv. Form elongate and narrow; the head broad and vertically inclined, a little narrowed behind the eyes, and broadly insunk in the thorax. The thorax, almost as broad as long and bulged at the side, is a little wider than the head (including the eyes) and much narrower than the elytra, which are long, very evensided and moderately convex. The coloration is tawny yellow with the eyes black and with a broad submedian pale greyish-purple stripe on each elytron, and the under surface of the abdomen greyish purple. The entire surface of the body, mouth-parts and limbs is clothed with fine fawn-grey pubescence, which has become matted with wrinkled effect on the thorax and elytra.

The *head*, broad and uniformly tawny yellow, rather flattened and finely punctulate, is narrow between the eyes, which are large, reniform and black. The antennae (both wanting) are inserted in front of the eyes between the notch and the fronto-clypeal suture; the clypeo-labral suture is straight; the tips of the palps and of the flattened bifid mandibles are black.

The *prothorax* (*pronotum*) is in front not as broad as the head including the eyes, but its widest part is broader than the head; its length is a little more than its greatest width; viewed from above it is subcordate, the sides being strongly rounded out behind the narrow and slightly constricted collar-like front; the front border is arcuate and is not raised, not marginate; the hind border is straight and marginate with a marked groove-constriction between the border and the disc. The moderately convex disc is somewhat flattened centrally and slightly depressed at the middle of the base; and there is also a diagonal depression at each side of the front. The sides of the pronotum are roughly triangular, strongly inflected and deep, bulged anteriorly and grooved posteriorly, the groove being parallel with the lateral continuation of the marginate hind border; and the lateral border is short and almost straight. The pronotal surface is closely punctulate and very pubescent.

The *prosternum* is transverse and has a minute pointed prosternal process between the bases of the front coxae. The *mesepisternum* is large and roughly square; the *mesosternum* is small and has a very small pointed process between the bases of the middle coxae. The *metepisternum* is transverse, and the *metasternum* is large and very convex.

The *scutellum* is relatively large, roughly triangular, rather obscurely punctulate, and a little depressed.

The *elytra* are fully four times as long as the thorax and about twice its breadth at the base; they are convex, the sides are strongly inflected and the apices are inflected over the end of the abdomen. The shoulders are prominent; and the

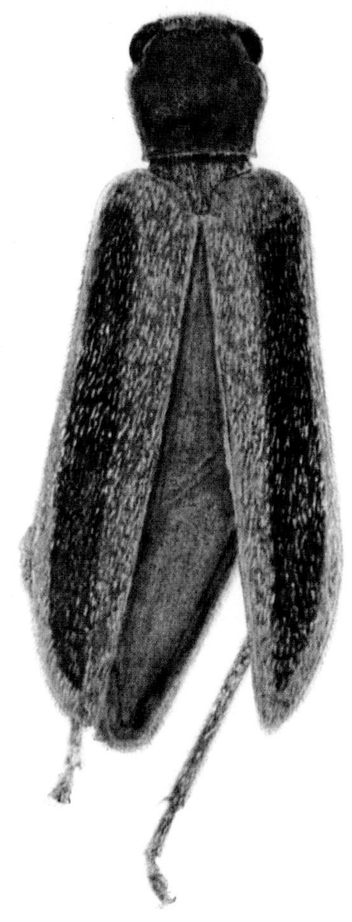

Oedemera vittata Oliv. × 16

outer sides, finely marginate, are parallel with the suture from the base to beyond the middle and from there gently rounded in apically. The elytral surface is obscurely punctulate and distinctly rugulose on the shoulder areas, and is coated with a fawn-grey pubescence; the coloration is tawny yellow with a broad violet-grey stripe on each elytron extending from the shoulder and tapered off at a short distance from the apex. This elytral stripe is not quite median, it is nearer the outer side.

The anterior coxal cavities are widely open behind and confluent. The front coxae are large, conical and contiguous and grooved on their outer sides; the hind coxae are transverse and marked with a strong groove, and the ante-coxal piece is very thin. The front and middle pairs of *legs* are wanting; the hind pair are moderately long; the femora are stout and arcuate, the tibiae have each two distal spines, the first or proximal segment of the tarsi is long, longer than the succeeding segments together, the penultimate segment is bilobed, and the claws are simple.

The sterna of the *abdomen* are very finely, rather obscurely punctulate and coated with a light yellowish pubescence. The anterior border of the first abdominal sternum has a small median pointed projection. The second sternum is longer than the others, which are about equal in length, except the sixth segment which is small, with a deep triangular notch. The first and second sterna are tawny yellow; the third, fourth and fifth are dark brown.

Length 10 mm.; breadth (across the middle of the elytra) 3 mm.
Hab. South America (Oliv.).
See Plate 51.

Family Rhipiphoridae

The following species of Rhipiphoridae is mentioned by Fabricius, in his published works, as having been described by him from a specimen or specimens in Dr Hunter's Collection:

Mordella sexmaculata *Syst. Ent.* p. 263, No. 4 (1775).

The above name is the original name as given by Fabricius, and the reference is to the work in which this species was first described.

The following species are the Yeats examples described by Fabricius and assumed to have been acquired by Hunter:

Mordella dimidiata	*Sp. Ins.* I, p. 332, No. 5 (1781).
limbata	*Ibid.* No. 6.

In the following pages the above-mentioned types are described under their *modern* names and in the order adopted in Junk and Schenkling's *Coleopterorum Catalogus.*

88. *Macrosiagon dimidiatum* (Fab.)

Coleopterorum Catalogus, pars 54 (E. Csiki, 1913), Rhipiphoridae, p. 11.

North America: New York, Florida, Missouri.

SYN. *Mordella dimidiata* Fab., *Sp. Ins.* I, p. 332, No. 5 (1781); *Mant. Ins.* I, p. 218, No. 9 (1787).
Ripiphorus dimidiatus Fab., *Ent. Syst.* I, 2, p. 112, No. 10 (1792); *Syst. Eleuth.* II, p. 120, No. 16 (1801); Oliv. *Ent.* III, 65, p. 8, pl. 1, figs. 8*a* and *b* (1795).
Macrosiagon dimidiatum (Fab.), Hentz, *Trans. Amer. Phil. Soc.* (2), III, 1830, p. 464.
Rhipiphorus dimidiatus (Fab.), *Gerst. Rhipiph. disp. Syst.* 1855, p. 21; Horn, *Trans. Amer. Ent. Soc.* V, 1874–76 (1875), pp. 121, 122.
Macrosiagon marginale J. Lec.

The type is stated by Fabricius to be in the Yeats Collection, and is supposed to have been acquired by Hunter.

Two specimens under label

'*Mord. dimidiata*
Fabr. pag. 332, No. 5'

in Cabinet B, drawer 14, answer the descriptions of this species given by Fabricius and Olivier, and correspond with

Olivier's figures of one of these Hunterian specimens. They have also been compared with modern examples in the British Museum and in the 'T. G. Bishop' Collection.

Description of Co-type, *Mordella dimidiata* Fab. The example is a *female*. Form elongate-oblong, rather spindle-shaped, narrowed towards the prominent head, widest about the middle, i.e. at the base of the thorax and elytra, and regularly continuous in outline viewed from above; the body laterally compressed, the sides of the thorax and abdomen deep; the elytra blade-like, with acuminate tips, and widely divergent. In side view the form is humpbacked (gibbous), the thorax being strongly arched in front; the short but vertically long head is markedly deflexed, and the short abdomen is terminally subtruncate. The surface mainly smooth, closely punctulate, and shining; coloration dull glossy brownish black excepting the elytra, which are partly reddish yellow, and the labial palps and basal segments of the antennae which are glossy reddish yellow.

The short but prominent *head* is wider than the front of the prothorax and is strongly constricted at the base, behind the eyes. It is vertically long and almond-shaped and is set on the narrowed front of the prothorax at right angles to it; but as the front portion of the prothorax is bent down, the head thus slopes obliquely inwards so that it rests upon the prosternum and close against the front coxae. The vertex (a little convex behind) is greatly elevated as a compressed dome above the level of the prothorax and perpendicular to it. The frons is flattened. The clypeus is transverse and (anteriorly) a little narrower than the frons; it is marked off from the frons at the fronto-genal articulations of the broad bases of the very long and stout sickle-shaped mandibles. The distal or labral border of the clypeus (clypeo-labral suture) is straight and is excavated along the edge, except in the middle, which forms a small raised lobe between the excavations. The labrum is an elongate narrowed lobe; it is glossy reddish brown and very finely punctulate, and it has a regular fringe of long light-coloured hairs. The maxillary palps are pitchy brown and clavate; the long bifid ligula and the labial palps are reddish yellow.

The eyes are ovate, not quite reniform, prominent and convex, finely faceted and glassy brown.

The antennae are inserted under the slight frontal projections of the small sockets immediately in front of the middle of the eyes; the sockets are open and continuous, between the clypeus and the lower portions of the eyes, as grooves extending on the genae. Only the long club-shaped first or basal segment and the short ring-like second segment of the antennae remain, and these segments are glossy light red. The coloration of the head is glossy dark pitchy brown and its surface is irregularly punctulate.

The *prothorax* is short, except the dorsal portion (pronotum), which is long, completely covering the mesonotum. Viewed from above, the *pronotum* is oblong and trapezoidal, roughly campaniform, as broad as the elytra at the base and gradually narrowed towards the front; it is mainly frontal and very convex, and in side view strongly arcuate and laterally prolonged at the anterior angles where it adjoins the front coxae. The posterior angles are blunt; the strongly inflected declivous sides overlap the mesothoracic pleura, and the free lateral borders are very arcuate. The narrow anterior or front margin (viewed from above) is almost straight, slightly sinuate and finely marginate; the posterior margin is sinuate and trilobed with two small lateral lobes (posterior angles) and with a large median triangular lobe which is hollowed and smooth without puncturation. The anterior edge of this hollow is thickened and raised as the broad end of a short median longitudinal rounded ridge which arises near the middle of the pronotum; the posterior edge of the hollow is emarginate, and within the hollow and between the elytra there is a very small *scutellum*. The pronotum is not marginate, except the narrow middle portion of its front border, which is very finely margined. The surface of the

Mordella dimidiata Fab. ♀ × 14

pronotum is irregularly punctulate, and the punctules are more or less contiguous posteriorly.

The propleura (*proepisterna* and *proepimera*) are greatly reduced, and not clearly marked off from the pronotum; the mesopleura and metapleura are large and irregularly punctulate. The free lateral edges (lateral or notopleural suture) of the pronotum curve forwards and downwards from the posterior angles above and in front of the large and convex mesopleura, which form the greater part of the deep sides of the thorax. The *mesepisternum* is more than twice the width of the *mesepimeron*, and these sclerites are clearly marked off by a distinct sinuous (mesopleural) suture. The roughly rectangular metapleuron overlaps the base of the abdomen, each side, and its lower or sternal border is almost straight; the *metepisternum* and *metepimeron* are triangular, and are marked off by an obliquely directed sinuous (metapleural) suture.

The *prosternum* is very small; the *mesosternum* is larger and it has a triangular process overlapping the middle coxae; the *metasternum* is large and very convex, its posterior border has a broad median triangular lobe overlying the hind coxae, and the apex of the lobe is finely emarginate and between the hind coxae. The thoracic sterna are dull glossy black, and irregularly punctulate.

The *elytra*, broadest at the base and narrowed to their tips, are blade-shaped with acuminate apices and widely separate (divaricate) from near the base; they are longer than the body but do not entirely cover the large wings, and are flattened above and finely marginate, and the sides are sharply inflected anteriorly; the outer borders are nearly straight, the inner borders are distinctly curved, the bases are rounded and the shoulders are prominent. The flattened surface of the elytra is uneven and punctulate, and the scattered oval punctules are more or less contiguous on the apical portions. The anterior portion of each elytron is reddish yellow, except the base, which has a narrow and irregular border of brownish black; the posterior or apical portions are dark brownish black.

The *legs* of this co-type are imperfect; they are long, dark brown, closely punctulate, covered with minute hairs, and have bifid claws. The front coxae are long, roughly conic, exserted and contiguous; the intermediate or middle coxae are long, semi-cylindrical and curved, and exserted; the hind coxae are small, rounded and distally contiguous, and with the transverse laminar portions deeply hollowed for the reception of the femora. The front legs have fairly stout femora and long trochanters, the tibiae are less stout and about half the length of the femora and have two distal spines or spurs; the tarsi are longer than the tibiae. The first tarsal segment is as long as the fifth and a little longer than the second, third and fourth taken together, these three segments being very small, especially the third and fourth, which together are about equal to the second. The middle legs are very long, the femora and tibiae are one-third longer than those of the front leg and proportionately stouter, and the trochanters are large; the tibiae have two long distal spurs; the tarsi are very long, longer than the femur and tibia together, and more than twice the length of the front tarsus. The first tarsal segment is the longest, its length is equal to that of the second, third and fourth taken together; the second is as long as the fifth, the third is shorter, and the fourth is the shortest segment.

The *abdomen* is glossy dark brownish black and irregularly punctulate; the scattered punctules bear fine short light yellow hairs. Being laterally compressed, the abdomen is narrow but deep; the first or basal segment is very long, several times longer than the two succeeding segments, which are very narrow; the remaining segments are defective in this specimen.

Length 7 mm.; breadth (across the base of the elytra) 2 mm.
Hab. North America (Fab.), not stated (Oliv.).
See Plate 52.

89. *Macrosiagon limbatum* (Fab.)

Coleopterorum Catalogus, pars 54 (E. Csiki, 1913), Rhipiphoridae, p. 13.

North America, Mexico, Guatemala, Costa Rica, Panama.

SYN. *Mordella limbata* Fab., *Sp. Ins.* I, p. 332, No. 6 (1781); *Mant. Ins.* I, p. 218, No. 10 (1787).
Ripiphorus limbatus Fab., *Ent. Syst.* I, 2, p. 112, No. 10 (1792); *Syst. Eleuth.* II, p. 121, No. 19 (1801); Oliv. *Ent.* III, 65, p. 6, pl. I, figs. 5*a* and *b* (1795).
Rhipiphorus limbatus (Fab.), *Gerst. Rhipiph. disp. Syst.* 1855, p. 30; Horn, *Trans. Amer. Ent. Soc.* V, 1874–76 (1875), pp. 122–125.

The example under label

'*Mord. limbata*
Fabr. pag. 332, No. 6'

in Cabinet B, drawer 14, is presumably the Yeats type acquired by Hunter; it answers the descriptions given by Fabricius, except that the '*scutellum ferrugineum*' is not visible. Olivier described this insect in Hunter's Collection and he mentions the scutellum, but he describes it as 'noir et très petit'; otherwise this type corresponds with Olivier's description and figures, and it has been compared with modern examples in the British Museum Collection.

Description of Type, *Mordella limbata* Fab. The example is a *female*. Form elongate and narrow and regularly continuous in outline, viewed from above; not bulging about the middle and the short but vertically long head not specially prominent, not wider than the front of the thorax, and the eyes not strongly convex. The body laterally compressed, the sides of the thorax and abdomen deep. The long narrow elytra blade-like, with the tips blunted, and widely divergent (from above the middle of the suture). In side view the form is humpbacked (gibbous), the thorax being so strongly arched in front that the head is completely inflexed beneath it and horizontal, in a line with the ventral surface of the body; and the short abdomen is abruptly rounded in, terminally sub-truncate in appearance. The head, thorax and abdomen, and the upper parts of the legs are mainly reddish yellow; the elytra and the antennae are mainly brownish black; the entire surface is punctulate and glossy.

The short *head* is vertically long, ovate in shape, as wide as the front of the prothorax and strongly constricted at the base, behind the eyes; it is broadly set on the narrowed front of the prothorax at right angles to it. As the front portion of the prothorax is strongly bent down, the head is strongly inflected and it rests upon the prosternum and against the front coxae. The glossy black rounded vertex, convex in front and behind, is considerably elevated as a dome, above the level of the prothorax and perpendicular to it. The frons is somewhat flattened, glossy reddish yellow and finely punctulate; the clypeus is transverse, anteriorly it is narrower than the frons, and it is marked off from the frons at the fronto-genal articulations of the bases of the long and moderately stout sickle-shaped mandibles; the distal or labral border of the clypeus (clypeo-labral suture) is almost straight (slightly sinuate) and is irregularly black. The labrum is an elongate lobe narrowed towards the front; it is glossy reddish yellow, finely punctulate and fringed with long yellow hairs. The palps (clavate) and the long ligula are reddish yellow.

The eyes are a little reniform, moderately convex, neither large nor prominent, and glassy brown.

The antennae are sub-serrate (the specimen being a female) and are inserted under the slight frontal projections of the small sockets immediately in front of the middle of the eyes; the first or basal segment, which is club-shaped, and the second segment, which is very small and rounded, are reddish yellow; the sub-serrate segments are brownish black.

The *prothorax* is short, except the dorsal portion (pronotum), which is long, completely covering the mesonotum. Viewed from above, the *pronotum* is oblong and campaniform, at the base it is as broad as the elytra and gradually narrowed towards the rounded front; it is mainly frontal and very convex, and in side view it is strongly arcuate and laterally prolonged at the anterior angles where it adjoins the front coxae; the posterior angles are sharply produced. The strongly inflected declivous sides overlap the mesothoracic pleura, and the free lateral borders are very arcuate. The narrow anterior or front margin (viewed from above) is almost straight and finely marginate; the posterior margin is deeply scalloped and trilobate with a large median and triangular scutellar lobe, the apex of which is blunt, and with two small lateral lobes (the produced posterior angles) which have acuminate apices.

The pronotum is not marginate, except the narrow middle portion of its front border, which is very finely marginate. The coloration of the pronotum is glossy reddish yellow around a large central and roughly hexagonal patch of blackish brown; the surface of the disc is punctulate and the scattered punctules are of irregular shape.

The propleura are greatly reduced and not clearly marked off from the pronotum; the mesopleura and metapleura are large and irregularly punctulate. The free lateral edges (lateral or notopleural suture) of the pronotum curve forwards and downwards from the posterior angles above and in front of the large and flattened mesopleura which form the greater part of the deep and flattened sides of the thorax; and the *mesepisternum* is black and very wide compared with the reddish yellow *mesepimeron*, which is a narrow strip clearly marked off by the distinct sinuous mesopleural suture; the roughly rectangular metapleuron overlaps the base of the abdomen, each side, and its lower or sternal border is straight; the glossy black and punctulate *metepisternum* and the reddish yellow and punctulate *metepimeron* are triangular and are marked off by an obliquely directed sinuous (metapleural) suture.

The *prosternum* is very small; the *mesosternum* is larger and has a long triangular process overlapping the middle coxae and its pointed tip extends a little beyond them; the *metasternum* is large and very convex, its posterior border has a broad median lobe overlying the bases of the hind coxae and the apex of the lobe is slightly emarginate and between the coxae. The thoracic sterna are irregularly punctulate and glossy reddish yellow, except the sides of the metasternum which are blackish brown.

The *scutellum* is not visible from above.

The *elytra*, broadest at the base and narrowed to their tips, are long and blade-shaped with blunt apices and widely separate (divaricate) from above the middle of the suture; they are longer than the body but do not entirely cover the large wings, and are flattened above and finely marginate, and the sides are sharply inflected anteriorly; the outer borders are nearly straight, except the apical portion which is gently incurved, the inner borders are curved outwards, the bases are rounded and the shoulders are moderately prominent. The flattened surface of the elytra is uneven and punctulate, and the scattered oval punctules are more or less contiguous on the apical parts. The coloration of the elytra is glossy blackish brown with a faint suffusion of red along the middle.

The *legs* of this type specimen are defective, with the exception of the right middle leg which is complete; they are long, closely punctulate, covered with minute hairs, and the claws are bifid. The front coxae are long, conic, exserted and contiguous; the intermediate or middle coxae are long, exserted, and deeply grooved on the outer side; the hind coxae are short, rounded, distally contiguous at the base, and the transverse portions are laminate with sharp posterior edges and hollowed for the reception of the femora. The femora of the front legs are moderately stout, the tibiae are less stout and about half the length of the femora. The middle legs are very long, the femora and the tibiae are about one-third longer and also stouter than those of the front legs; the tarsi are very long, as long as the femur and tibia together or about half the length of the whole leg; the first tarsal segment is the longest, the second, third and fourth segments are progressively shorter, the fifth tarsal segment is as long as the second but more slender. The hind legs also are very long; the femora are about the same size as those of the middle legs, but the tibiae are much longer, and the first tarsal segment is longer than that of the middle tarsus. All the trochanters are short, and the tibiae have each two fairly long distal spines or spurs.

The coxae are glossy reddish yellow and also the femora, except their distal ends which are black; the front and middle tibiae are black, the proximal half of the hind tibiae is glossy reddish yellow and the distal half is black; the front and middle tarsi are black and also the hind tarsi, except the proximal end of the first segment which is reddish yellow.

The short *abdomen* is very narrow but deep, the sides being greatly compressed and very flat; it is glossy reddish yellow, except the terminal segment which is black, and its surface is irregularly and not closely punctulate. The first or basal segment is very long, as long as the other four segments together; the lower or sternal portions of the second, third and fourth segments are equally narrow and their upper or tergal portions are wider, and the fifth or terminal segment is small and black. The upper part of the fourth abdominal segment is abruptly bent down, consequently the end of the abdomen (viewed from the side) is subtruncate in outline.

Length 6½ mm.; breadth (across the base of the elytra) 1½ mm.
Hab. Not stated (Fab. and Oliv.).

90. *Macrosiagon pectinatum* (Fab.)

Coleopterorum Catalogus, pars 54 (E. Csiki, 1913), Rhipiphoridae, p. 14.

North America (U.S.), Mexico, Guatemala.

SYN. *Mordella sexmaculata* Fab., *Syst. Ent.* p. 263, No. 4 (1775); *Sp. Ins.* I, p. 332, No. 4 (1781); *Mant. Ins.* I, p. 218, No. 7 (1787).
Ripiphorus sexmaculatus Fab., *Ent. Syst.* I, 2, p. 111, No. 8 (1792); *Syst. Eleuth.* II, p. 120, No. 12 (1801); Oliv. *Ent.* III, 65, p. 7, pl. 1, fig. 6 (1795).
Mordella pectinata Fab., *Syst. Ent.* p. 263, No. 3 (1775); *Sp. Ins.* I, p. 332, No. 3 (1781); *Mant. Ins.* I, p. 218, No. 3 (1787).
Emenadia pectinatum (Fab.).
Ripiphorus pectinatus Fab., *Ent. Syst.* I, 2, p. 111, No. 4 (1792); *Syst. Eleuth.* II, p. 119, No. 5 (1801).
Ripiphorus humeratus Fab., *Syst. Eleuth.* II, p. 119, No. 8 (1801).
R. tristis Fab., *Syst. Eleuth.* II, p. 119, No. 9 (1801).
R. nigricornis Fab., *Syst. Eleuth.* II, p. 119, No. 10 (1801).
R. ventralis Fab., *Syst. Eleuth.* II, p. 120, No. 13 (1801).
Rhipiphorus ambiguum Melsh.
R. dubium Melsh.
R. faciatum Melsh.
R. nigrum Melsh.

Two specimens, one of which is defective, under label

'*Mord.* 6-*maculata*
Fabr. pag. 332, No. 4'

in Cabinet B, drawer 14, are apparently the examples on which Fabricius founded this species, they answer his original description of *Mordella sexmaculata*. This species was later described by Fabricius as *Ripiphorus* 6-*maculatus* from an example stated to be in the Banks Collection; but, as indicated by Olivier and as pointed out by Dr K. G. Blair,[1] the descriptions of both are the same, and no such type exists

[1] 'Further Notes on the Fabrician Types of Heteromera (Coleoptera) in the Banks Collection' by K. G. Blair (*Annals and Magazine of Natural History*, Vol. v, Ninth Series, February 1920).

in the Banks Collection. The Hunterian co-type described below corresponds with the description and figure given by Olivier; it has also been compared with modern examples in the British Museum Collection and found to agree.

Description of Co-type, *Mordella sexmaculata* Fab. Form elongate-oblong, narrowed towards the head, widest and rounded at the shoulders and regularly continuous in outline; the body laterally compressed, the sides of the thorax and abdomen deep, and with widely divergent blade-like elytra. In side view the form is gibbous, the thorax being strongly arched in front, and the vertical head is markedly deflexed. Surface mainly smooth and coloration dark brown with the thorax (pronotum) reddish, the elytra reddish yellow and black, the head (vertex) and the bases of the antennae and the palps reddish yellow; the legs uniform reddish black.

The *head* is comparatively small, deflexed beneath the thorax and partly overlapping the front coxae, strongly constricted at the neck, vertically long and obovate; it is flattened in front and convex towards the vertex with a median vertical suture (ridge) extending from the ridge-like crown to the occipital foramen. The surface of the head is closely punctulate, except about the crown, and the punctules are largely confluent with rugulose effect on the fronto-clypeus and on the occipital area. The clypeus is transverse, narrowed in front and straight. The labrum is short and semi-lunar with a fringe of golden brown hairs; and the maxillary palps, which have lobate tips, are orange yellow. The eyes are relatively small, oval, convex and entire. The antennae are inserted in large deep foveae behind small protuberances, very near to and in front of the eyes, opposite the middle of the ocular margin; the basal segments of the antennae (otherwise imperfect) are clavate and orange yellow.

The *prothorax* viewed from above is oblong and trapezoidal, somewhat campaniform, at the base as broad as the elytra and gradually narrowed towards the front which is narrower than the head; the surface is irregularly and closely punctulate, the punctules largely confluent with imbricate effect. The *pronotum* is strongly arcuate, very convex, mainly frontal and not marginate; it is laterally prolonged at the front angles where it meets the front coxae; the posterior angles are blunt; the declivous sides overlap the mesothoracic pleura and the lateral borders are clearly marked and arcuate; the anterior or front margin (viewed from above) is straight, the posterior margin is sinuate with a broad median and triangular lobe which is slightly raised and overlying the scutellum, which is not visible. Upon the pronotum there is a slight median longitudinal ridge, which extends from the anterior margin to a small median fovea at the apex of the triangular scutellar lobe; between this ridge and each posterior angle the pronotal surface is hollowed. The mesopleura and metapleura are large and closely punctulate; the broad *mesepisternum* is marked off from the very narrow *mesepimeron* by a distinct sinuous suture; the rectangular metapleuron overlaps the base of the abdomen, and its lower or sternal border is straight; the *metepisternum* and *metepimeron* are triangular and marked off by an obliquely directed sinuous suture. The *prosternum* is very small; the *mesosternum* has a separate process extending between the middle coxae, and the large *metasternum* has a small median angular lobe, overlying the hind coxae.

The *elytra*, narrowed posteriorly and blade-shaped with acuminate apices, are widely separate (divaricate) from near the base, marginate, and do not entirely cover the large wings; the outer borders of the elytra are nearly straight, the inner borders are curved. The elytral surface, closely punctulate with oval punctules, is flattened above, rounded on the inflexed sides, and the shoulder callus is prominent. The colour pattern of the elytra is reddish yellow with the bases and the tips brownish black; there is also a large irregular brownish black spot about the middle of each elytron and adjoining the outer margin, and the inner margins are narrowly bordered with brownish black.

The *legs* are uniform reddish black, long and slender; the

The *prosternum* is short and transverse with a large triangular sharp-pointed prosternal process between the anterior or front coxae. The *metasternum* is long and very convex; it is closely and finely punctulate with imbricate effect and is largely coated with short and recumbent yellowish hairs. A faintly impressed median longitudinal line extends from the anterior border to the posterior border, there ending in a small median sulcus between two small lobe-like projections. The posterior border is very sinuate about the coxae.

The *scutellum* is dark reddish brown; it is elongate and triangular with the sides incurved and the apex blunt, its surface is impunctate but not quite smooth and is marked by a slight but distinct median longitudinal raised line.

At each side of the mesonotum the ear-like auxiliary parts of the *elytra* are exposed to view. The elytra are longer than the body, extending well beyond the end of the abdomen, and they are narrowed towards the apices; the length is two and a half times the breadth across the base, which is nearly twice as broad as the prothorax, each elytron being 4½ mm. at the base and less than 3 mm. near the apex. The elytra are vertically raised and very convex about the base, with prominent rounded shoulders, but rather flattened from behind the shoulders towards the apices. The base is oblique from the suture to the axilla, and widely rounded at the shoulder angles; the inflexed sides are vertical and very narrow from the middle towards the apex, and (viewed from above) are distinctly sinuous; the apices are abruptly rounded in and have slightly projecting sutural tips. From the middle of the suture the elytra are apically divergent and therefore dehiscent; and they are finely marginate, except about the middle of the base.

On each elytron there are four longitudinal raised lines, like ribs. The first or innermost rib extends from the base to the apex and closely adjoins the suture, except near the base, where it diverges outwards. The second rib is in great part parallel with the first; but it does not reach the apex, anteriorly it diverges outwards to the middle of the base and posteriorly it merges in the first rib towards the apex; it is also connected with the first rib by three short cross-ribs, one posterior, one near the middle and one near the base where both ribs diverge outwards. The third or middle rib is the shortest, it extends only as far as the apical third of the elytron, and at both ends it is joined to the second rib. The fourth or outermost rib extends from the outer side of the shoulder callus and becomes obsolete towards the elytral apex; it forms the upper boundary of the inflexed side, and it bifurcates anteriorly, the branch being within the wider portion of the inflexed side. These ribs are not exactly similar on both elytra, and under the lens they have a ramose appearance with irregular fine branches.

The coloration of the elytra is uniformly testaceous (yellowish red); the elytral surface is closely and irregularly punctulate and very rugulose with ramose or branching effect, except upon the basal areas at each side of the suture where the ribs diverge, these parts being smooth and faintly punctulate. There is a small black and obliquely transverse area, closely set with fine longitudinal ridges, on each elytron at the middle of the base.

The anterior coxal cavities are large, confluent and widely open behind, and the posterior coxal cavities are prominent acetabula.

The *legs* are long, dark chestnut brown and punctate with short yellowish hairs; the front and middle pairs are close together. The front and middle coxae are large, conic, exserted and contiguous; the hind coxae are transverse, posteriorly grooved, and contiguous. The femora are moderately stout. The tibiae have each two short distal spurs, but those of the hind tibiae are lobe-like; and the hind tibiae are very arcuate. The tarsi are somewhat flattened and long; the front and middle tarsi are slightly longer than the tibiae, and are five-segmented; the first or basal segment is as long as the fifth, the other segments are progressively shorter; the first four

PLATE 53

Mordella sexmaculata Fab. × 12
and side view

front coxae are large and exserted, conical and contiguous, and the intermediate or middle coxae are similar but apart; the hind coxae are small, transverse and almost contiguous and their transverse laminar portions are hollowed for the reception of the femora; the hind femora are broad and flattened; the middle and hind tibiae have prominent distal spurs; the front and middle tarsi are five-segmented, the hind tarsi four-segmented; the middle tarsi are longer than the tibiae, the first or basal tarsal segment being nearly as long as the tibia; the claws are large and bifid.

The *abdomen* is uniformly dark chocolate brown, with markedly compressed sides, and there are four visible segments; the basal segment is very long, twice the length of the succeeding one which is longer than the third, and the fourth segment is very narrow; the sterna are very deep. The surface of the abdomen is closely punctulate; the punctules of the sternal surface are very fine and bear short golden hairs.

Length 8 mm.; breadth (across the base of the thorax) 3 mm.
Hab. America (Fab. and Oliv.).
See Plate 53.

Family Meloidae

The following are the species of Meloidae mentioned by Fabricius, in his published works, as having been described by him from specimens in Dr Hunter's Collection:

Lytta vittata	*Syst. Ent.* p. 260, No. 3 (1775).
atrata	*Ibid.* p. 260, No. 4.
Lagria marginata	*Sp. Ins.* I, p. 159, No. 5 (1781).

The above names are the original names as given by Fabricius, and the references are to the works in which these species were first described.

The following species were described by Olivier from examples in Dr Hunter's Collection:

Cantharis testacea	*Ent.* III, 46, p. 7, pl. 2, fig. 11 (1795).
Mylabris ruficollis	*Ibid.* 47, pp. 14–19, pl. 2, fig. 17.

Sub-family LYTTINAE

Tribe Mylabrini

91. *Eletica testacea* (Oliv.)

Coleopterorum Catalogus, pars 69 (F. Borchmann, 1917), Meloidae, etc. p. 66.

Senegal

SYN. *Cantharis testacea* Oliv., *Ent.* III, 46, p. 7, pl. 2, fig. 11 (1795).

The insect in Cabinet B, drawer 13, with attached label '*Lytta testacea* Oliv.' (the handwriting presumably Olivier's) is the *Eletica testacea* of Borchmann and the Olivier type of this species; it answers Olivier's description closely and corresponds with his illustration, which represents it with one side partly exposed, exactly as it is pinned.

In his description, Olivier refers to fig. 10, but fig. 11 is obviously *testacea*, as it clearly shows the raised longitudinal lines on the elytra; and on each elytron there are four such lines, not two as stated by him.

Lytta testacea is a Fabrician insect, described by Olivier as *Cantharis melophtalmos*.

Description of Type, *Cantharis testacea* Oliv. The specimen is a *male*. Form elongate and strongly convex, the conspicuous vertical head and the prothorax about equal in width but much narrower than the elytra, which are long, narrowed posteriorly and apically dehiscent, with the sides inflexed and slightly sinuate, and with four longitudinal and irregular raised rib-like lines upon each elytron. The rounded pronotum constricted and thickened (collar-like) in front and a little constricted behind; the legs long. Coloration yellowish brown (testaceous), except the basal segments of the antennae, the pronotum and the scutellum, which are brownish black; the underparts of the body and the legs dark chestnut brown and more or less coated with moderately long recumbent yellowish hairs.

The *head* is vertically placed and vertically long, it is connected with the thorax by a short and narrow neck, frontally and also laterally it is broadest across the temples, its greatest breadth being 4 mm., or nearly that of the prothorax, which measures 5 mm. The vertex (epicranium) is testaceous (yellowish brown) and it is high and rounded; it is deeply marked by a median epicranial furrow which extends from the occipital foramen to the frons; the surface of the vertex is thinly punctulate and on each side above the eye (between the vertex and the temple) there is a short horizontal furrow. The frons is brownish black, coarsely punctate and rugulose; it is long and narrow, widening out immediately below or in front of each eye into two triangular parts distinctly marked off and bearing the raised antennal insertions. Above the eyes, the frons is clearly defined between the two divergent and incurved arms of the median epicranial suture; and it is a little convex towards the clypeus. The fronto-clypeal suture is arcuate. The clypeus is transverse, partly black, partly yellowish red and smooth. The clypeo-labral suture is straight, and the labrum is a transverse lobe fringed with long yellow hairs.

The eyes, which are horizontally placed, extend round the sides of the head and encroach considerably on the frons; they are large and reniform, very convex, finely faceted and yellowish within their margins. The thick mandibles appear to be bidenticulate; the maxillary palps are exserted and the distal segment is short and lobate.

The eleven-segmented antennae are not long, reaching just a little beyond the thorax; they are brown, except the first two basal segments, which are black, and they are inserted immediately in front of (or below) the anterior parts of the eyes. The first antennal segment is the longest and it is club-shaped, the second segment is the shortest and it is obconical; the third segment is shorter than the first and the remaining segments are progressively shorter, somewhat flattened and also slightly serrate.

The *prothorax* is much narrower than the elytra, its breadth is 5 mm. or about one-half the width of the elytra, which measure 9 mm. across the base; it is constricted, very narrow and collar-like at its junction with the neck. The *pronotum* is very convex and abruptly declivous before the front; viewed from above it is very rounded and bulging over the sides, greatly constricted in front, there forming a narrow collar at the neck, and a little constricted before the margin of the broad base, which is more than twice the width of the front. The sides are vertical, deep and angular and are marked off by a lateral horizontal furrow from the bulged-out disc which hides them from upper view. The disc is divided into a double elevation by a median longitudinal linear furrow. This furrow is interrupted in the middle; its anterior part becomes obsolete near the front, and its posterior part ends in conjunction with two curved lines which border the fovea on each side of it, between the double elevation and the margin of the base. The pronotum is marginate all round; the basal portion of the margin is vertically raised and prominent, partly double and slightly sinuate. The glossy black surface of the pronotum is thinly punctulate over the uneven disc, closely punctulate in front and also rugulose on the collar, and the steep sides are closely punctulate.

The *mesothorax* is largely exposed. The *mesonotum* is subquadrate; the front and sides are a little incurved and the base is prolonged into the scutellum. The front part of the mesonotum is hollowed, it is also divided by a deep median sulcus; and a thin raised line, continuous with the sulcus, marks the slightly convex basal part. The surface of the mesonotum is irregularly punctulate, rugulose and blackish brown. The *proepisterna* are small triangular parts adjoining the gula on each side. The *mesepisterna* are long and finely rugulose; the *mesepimera* are obliquely subquadrate, irregularly punctulate, and marked by a central oval elevation. The *metepisterna* are long and narrow; the surface is uneven and closely punctulate with imbricate effect.

Cantharis testacea Oliv. ♂ × 4
and side view

segments are deeply notched (bilobed), the fifth segment is thinner than the others and subcylindrical; the claws are simple, and each claw has an accessory lateral claw-like appendage of the same length but thinner. All the tarsal segments are thickly fringed at the sides and beneath with short bright yellow hairs. The hind tarsi are four-segmented and shorter than the tibiae.

The *abdomen* has five visible sterna, which are about equal in length; the first or basal sternum is the broadest, the last or fifth is narrowed and its posterior border is a little arcuate. The surface of the abdominal sterna is closely and finely punctulate with imbricate effect, and thinly coated with fine yellowish recumbent hairs.

Length 33 mm.; breadth (across the shoulders of the elytra) 9 mm.
Hab. not stated by Olivier.
See Plate 54.

Tribe Lyttini

92. *Epicauta pennsylvanica* (De Geer)

Coleopterorum Catalogus, pars 69 (F. Borchmann, 1917), Meloidae, etc. p. 79.

North America, Missouri, Pennsylvania.

SYN. *Meloë pennsylvanica* De Geer, *Mém. Hist. Ins.* v, p. 16, pl. 13, fig. 1 (1775).
Lytta atrata Fab., *Syst. Ent.* p. 260, No. 4 (1775); *Sp. Ins.* I, p. 329, No. 7 (1781); *Mant. Ins.* I, p. 216, No. 8 (1787); *Ent. Syst.* I, 2, p. 86, No. 12 (1792); *Syst. Eleuth.* II, p. 79, No. 19 (1801).
Cantharis atrata (Fab.), Oliv. *Ent.* III, 46, p. 17, pl. 2, fig. 19 (1795).
Epicauta coracina Ill. Carolina.
E. morio J. Lec. Carolina.
E. nigra Woodh. Mexico.
E. potosina Dugès.

There are four insects under label

'*Lytt. atrata*
Fabr. pag. 329, No. 7'

in Cabinet B, drawer 13; and two of these belong to a closely related genus; they have been identified as *Macrobasis cinerea* (Fab.) syn. *unicolor* Kby. The other two specimens are co-types of *atrata* Fab.; they correspond with the descriptions of this species given by Fabricius and Olivier and with Olivier's figure, and they closely match modern examples in the British Museum Collection.

Description of Co-type, *Lytta atrata* Fab. Form elongate and narrow. The large head transverse and vertical, abruptly constricted behind and broadly set against the small thorax, which is narrower than the head and much narrower than the elytra, which are long, almost parallel-sided and divaricate at their broadly rounded apices. The antennae moderately long and filiform; the legs long. Uniformly deep reddish black, with the head, antennae, legs and underparts of the body distinctly glossy.

The *head* is connected with the thorax by a very short neck, the upper part of which is visible and lozenge-shaped, with a few coarse punctures bearing short and fine light golden hairs. Vertically placed and vertically long, the head is large and transverse (viewed from above), broader than the prothorax and fully two-thirds the width of the elytra; it is hollowed occipitally and convex in front. The vertex (epicranium) is broad, high and rounded and is marked by a rather faint median longitudinal line (median or epicranial suture), which extends from the occiput over the vertex and the frons but which becomes obsolete before the arcuate fronto-clypeal suture. The clypeus is transverse, narrowed towards the front, and has a small red spot at each side just above the straight clypeo-labral suture; the labrum is notched in front and fringed with long golden hairs. The surface of the head is uniformly dull glossy black, irregularly and closely punctulate and set with very short fine recumbent hairs.

The palps are black and their distal segments are securiform. The eyes are golden green, reniform and deeply emarginate above the antennal insertions. The dull glossy

Lytta atrata Fab. × 11

reddish black and filiform antennae (one imperfect) are inserted in front of the eyes, at the sides of the frons, between the eye-notch and the fronto-clypeal suture. The first or basal segment is long, clavate and distinctly stouter than the other antennal segments; the second segment is the shortest, the third is nearly as long as the first, and the remaining eight segments are progressively thinner towards the tip of the antenna.

The *prothorax* is narrower than the head and much narrower than the elytra, and its length is greater than its breadth; viewed from above it is oblong with the lateral borders rounded off and with wedge-shaped sides. The *pronotum* is saddle-shaped and the inflexed deep sides overlap the bases of the front coxae; it is oblong above, parallel-sided from the base to beyond the middle, and gradually narrowed anteriorly; the front margin is slightly excavate, the hind margin is straight, and the anterior and posterior angles are rounded off; the disc is moderately convex, and its surface is uneven with shallow depressions, there being a median basal fovea with a small round depression on each side of it and anteriorly a large median shallow depression. A deeply marked median longitudinal suture extends over the disc from the basal fovea to near the front margin. The pronotum is uniformly dull reddish black, finely marginate, and its surface is closely punctulate; the punctules, which are confluent on the basal portion of the disc, bear very short and fine recumbent hairs.

The *mesepimera* and the *metepisterna* are largely exposed, not entirely covered by the inflexed sides of the elytra, and the metepisterna are hollowed. The *prosternum* is small; the *mesosternum* is much larger and has a small pointed process between the bases of the middle coxae. The *metasternum* is large (long) and strongly convex, and it has a median process with a small notch between the hind coxae. The surfaces of the pleura and sterna are marked with fine and confluent punctulation and clothed with fine recumbent hairs.

The *scutellum* is small and triangular, with the apex rounded; its surface is punctulate and set with fine recumbent hairs.

The *elytra* are long and narrow, longer than the body, reaching beyond the end of the abdomen, and the breadth across their bases is fully one-third greater than that of the base of the thorax; they are very convex and considerably inflexed at the sides, which are almost perfectly parallel, the very slight sinuation behind the shoulders being scarcely perceptible, and the evenly rounded apices are a little divaricate at the suture. The base of the elytra is angular, obliquely curved against the scutellum from the suture to the middle and broadly rounded in front of the shoulders, which are raised and prominent. The elytra are finely marginate all round, uniformly dull reddish black, and the surface is very closely and finely punctulate, with imbricate effect, and covered with fine recumbent hairs. Upon each elytron there are two very faint subparallel lines, one along the middle and the other between the middle and the outer border.

The *legs* are long and uniformly dull glossy reddish black; but the anterior femora have, upon the inner borders and about the middle, a small triangular patch of brilliant golden recumbent hairs. The front and middle coxae are long, roughly conic in shape and exserted; the front coxae are contiguous and the middle coxae nearly so; the hind coxae are prominent, transverse and near together. The femora are stout, the tibiae have each two distal spines; the front and middle tarsi are five-segmented, the hind tarsi are four-segmented, and the claws are double.

The six visible sterna of the *abdomen* are of approximately equal length; the sternal surface is very closely punctulate, with imbricate effect, and thinly coated with fine recumbent hairs.

Length 10 mm.; breadth (across the elytra) 3 mm.

Hab. America (Fab.), North America, in Carolina and Pennsylvania (Oliv.).

See Plate 55.

93. *Epicauta vittata* (Fab.)

Coleopterorum Catalogus, pars 69 (F. Borchmann, 1917), Meloidae, etc. p. 85.

North America.

Syn. *Lytta vittata* Fab., *Syst. Ent.* p. 260, No. 3 (1775); *Sp. Ins.* I, p. 329, No. 6 (1781); *Mant. Ins.* I, p. 216, No. 7 (1787); *Ent. Syst.* I, 2, p. 86, No. 11 (1792); *Syst. Eleuth.* II, p. 79, No. 18 (1801).

Cantharis vittata (Fab.), Oliv. *Ent.* III, 46, p. 13, pl. 1, fig. 3 (1795).

There are three insects under label

'*Lytt. vittata*
Fabr. pag. 329, No. 6'

in Cabinet B, drawer 13; one of these has been identified as *Epicauta fumosa* Germ., and the other two are evidently co-types of *vittata* Fab., they answer the descriptions of this species given by Fabricius and Olivier and closely agree with recent examples in the British Museum Collection.

Olivier's figure of *vittata* is a poor representation, more especially of the thorax, the three markings upon it not being properly indicated. These thoracic markings are described as lines by Fabricius and also by Olivier; but only the middle one is narrow and linear in the co-types, the other two (dorsi-lateral) are irregular and have the form of roughly triangular patches posteriorly.

Description of Co-type, *Lytta vittata* Fab. Form elongate and narrow. The large head vertical, abruptly constricted behind and broadly set against the small thorax, which is narrower than the head and also narrower than the elytra, which are long and almost parallel-sided and divaricate at their rounded apices. The antennae and the legs long. Coloration of the head yellowish orange-red with two dark brown spots; the thorax dark brown with three reddish yellow markings; the elytra dark brown bordered with reddish yellow, and with a reddish yellow median vitta confluent with the border at the apex of each elytron; the proximal or basal parts of the femora reddish yellow. The entire surface of the body and legs closely set with very short and fine recumbent golden hairs, longer on the underparts.

The *head*, connected with the thorax by a very short and narrow neck, is vertically placed and vertically long; it is large, broader than the prothorax and more than two-thirds of the width of the elytra; and it is dilated above the level of the fronto-clypeal suture. The vertex (epicranium) is wide, high and rounded, and it is divided by a median longitudinal linear furrow (median or epicranial suture) which extends from the occiput over the vertex and the frons to the arcuate fronto-clypeal suture. The semicircular clypeus is narrower than the frons, the clypeo-labral suture is nearly straight; and the labrum is as broad as the clypeus and deeply notched in front. The surface of the head is yellowish orange-red, closely punctulate and covered with short recumbent golden yellow hairs; on the clypeus and labrum the puncturation is coarser and less close, and the hairs are longer, forming fringes. Upon the frontal part of the vertex, above the level of the eyes, there are two large dark brown spots, reniform in shape and close together, one at each side of the median epicranial suture. The palps are dark brown and their distal segments are securiform. The eyes are transverse and biemarginate; the corneal surface is finely granulose. The antennae (imperfect in this specimen) are inserted in front of the eyes at the sides of the frons, between the upper eye-notch and the fronto-clypeal suture.

The *prothorax* is much narrower than the head and the elytra, and its length is a little more than its breadth; viewed from above it is oblong, in side view it is wedge-shaped. The *pronotum* is saddle-shaped with deep inflexed sides reaching to the bases of the front coxae; it is oblong above and subparallel, being parallel-sided from the base to beyond the middle and gradually narrowed anteriorly. On the convex

Lytta vittata Fab. × 8

disc there are three depressions (one median and two lateral) between the middle and the front; and at the base there is a median triangular fovea. A well-marked median longitudinal suture extends from the basal fovea over the convexity of the disc, becoming obsolete anteriorly. The pronotum is marginate, and its surface is punctulate, the punctules bearing light golden hairs; the colour pattern is blackish brown, with a narrow strip of reddish yellow about the suture and a yellowish red patch at each side posteriorly.

The *mesepimera* and the *metepisterna* are exposed, not covered by the inflexed side portions of the elytra; and the metepisterna are hollowed. The *prosternum* is small; the *mesosternum* is larger, with a small pointed process between the bases of the middle coxae; the *metasternum* is large (long) and very convex, with a median transverse lobe between the hind coxae.

The *scutellum* is small and triangular, with incurved sides and the apex rounded; the surface is punctulate and covered with light golden recumbent hairs.

The *elytra* are long and narrow, longer than the body, reaching well beyond the tip of the abdomen; and the width across the base is fully one-third more than that of the thorax. They are regularly convex and narrowly inflexed at the sides, which (viewed from above) are a little sinuate, almost parallel, there being a slight constriction behind the shoulders; and the evenly rounded apices are divaricate at the suture. The base of the elytra is angular, obliquely curved against the scutellum from the suture to the middle and broadly rounded in front of the raised and prominent shoulders. The suture is finely marginate, but the margin becomes obsolete towards the apex. The elytral surface is closely and finely punctulate, and thinly covered with short and fine recumbent yellow hairs. The coloration of the elytra is dark brown with a very regular narrow and complete border of reddish yellow, and with a longitudinal median stripe of reddish yellow extending from the middle of the base to the apex, there turning in towards the suture and forming a loop junction with the sutural portion of the reddish yellow border. Along the inner or sutural side the reddish yellow border is narrower than on the outer side and around the apex; and the median stripe is twice the breadth of the sutural border and a little broader than the outer border. A distinct impressed line extends from the middle of each shoulder along the median stripe, becoming obsolete about the apical third of each elytron.

The anterior and the middle coxal cavities are large, confluent and open behind. The *legs* have the basal parts of the femora reddish yellow; the front and middle coxae are long, somewhat conical, exserted, close together and distally contiguous; the hind coxae are transverse, grooved, and near together; the front femora are distally notched or grooved; the front and middle tibiae have each two distal spines; the claws are double.

The sterna of the *abdomen* are coated with light yellowish recumbent hairs, and the punctulation of the surface is linear with imbricate effect.

Length 15 mm.; breadth (across the middle of the elytra) $4\frac{1}{2}$ mm.
Hab. America (Fab.).
See Plate 56.

94. *Tetraonyx quadrimaculatus* (Fab.) var. *bimaculatus* Klug.

Coleopterorum Catalogus, pars 69 (F. Borchmann, 1917), Meloidae, etc. p. 114.

Mexico, Brazil.

SYN. *Mylabris ruficollis* Oliv., *Ent.* III, 47, pp. 14–19, pl. 2, fig. 17 (1795).
Apalus 4-maculatus Fab., *Ent. Syst.* I, 2, p. 50, No. 2 (1792); *Syst. Eleuth.* II, p. 25, No. 2 (1801).

Two specimens in Cabinet B, drawer 13, with attached label ('*Mylabris ruficollis* Oliv.') presumably in Olivier's handwriting, are probably co-types of this species; they correspond closely with the description and figure given by Olivier. Dr K. G. Blair examined the co-type here described and he pointed out that the shape of the head is that of the South American form.

Description of Co-type, *Mylabris ruficollis* Oliv. Form oblong and moderately convex; the elytra fully twice as long as broad, slightly expanded behind the middle, almost parallel-sided and much broader than the transverse thorax, which is not much wider than the large vertical head. The eyes reniform, very convex, fawn-coloured, and moderately conspicuous. The scutellum comparatively large; the epipleura extremely narrow, and the legs long. General coloration black with tawny yellow; the head and antennae black, the thorax and scutellum tawny yellow; the elytra black with a broad irregular transverse band of tawny yellow across the middle, and with a continuation of the same colour on the suture between the scutellum and the band; the legs black with the femora three-quarters tawny yellow; the entire upper surface coated with short and fine hairs, the under surface partly black and more or less coated with light tawny yellow hairs.

The *head* is uniformly black, it is vertically placed and vertically long, nearly as broad as the thorax and constricted at the short neck. The vertex (epicranium) is high and rounded; the frons is convex and not greatly narrowed between the eyes, which are wide apart, and it overhangs the hollowed clypeus which is narrower and transverse. The clypeo-labral suture is straight and the labrum is rounded and emarginate in front. The distal segment of the exserted maxillary palps is clavate and flattened. The gena is narrowly transverse; and the very convex, reniform and fawn-coloured eyes, being placed obliquely upwards towards the vertex, do not encroach to any great extent on the frons. The antennae are almost entirely wanting in this type specimen, only two basal segments remain, the first short and conic and the second broadly clavate. The antennal insertions in front of the eyes are situated below the middle of each eye, opposite the lower part of the notch, and do not project prominently. The surface of the head is closely punctate, and about the frons it is also rugulose.

The *pronotum* is uniformly tawny yellow, irregularly punctulate, and coated with light yellowish recumbent hairs; it is about one-half the width of the elytra at the base. The shape of the pronotum is transverse, it is narrowed in front, with a narrow collar-like constriction about the neck; the anterior angles are rounded, the sides (viewed from above) are evenly curved, and it is broadest at the base which is a little sinuous, with the posterior angles distinctly projecting, rounded off and hollowed within the angle. The margin of the base is interrupted about the middle by a slight overlying scutellar lobe; the raised disc is regularly convex, and laterally it forms a thickened ridge overhanging the sides (hidden from view above) which are deep and hollowed, with the lower border angular and finely marginate. There is a large pit or fovea on each side anteriorly. The pronotum is marginate on the collar, and the margin of the angular sides is continuous with the interrupted margin of the base and is noticeably thickened around the posterior angles.

The roughly triangular *mesepisterna* and the obliquely subquadrate *mesepimera* are entirely exposed to view, their surface is black and covered with short light-coloured recumbent hairs; the *mesosternum* is transverse, with a short median process between the contiguous middle coxae. The *metepisterna* are long and triangular, black and coated with short yellowish recumbent hairs. The *metasternum* is long, very convex, black, and thickly covered with short and recumbent yellowish hairs.

The *scutellum* is tawny yellow, comparatively large and conspicuous; it is long and triangular with the sides incurved and the apex blunt; its surface is punctulate and closely invested with light yellowish recumbent hairs.

The *elytra* are regularly convex and moderately long, fully

Mylabris ruficollis Oliv. × 11

twice as long as broad; and the width across the base is nearly twice that of the prothorax. The base is angular, obliquely curved against the scutellum from the suture to the middle, and broadly rounded in front of the raised and prominent shoulders. The narrowly inflexed sides (viewed from above) are subparallel, slightly sinuate (a little constricted behind the shoulders and slightly expanded about the middle) and evenly rounded apically to the suture. The apices appear to be only slightly divergent at the suture; the sides and the suture are finely marginate, but the side and sutural margins become obsolete towards the apex. The surface of the elytra is finely punctulate, the punctules bearing short recumbent black and yellowish hairs; and the coloration is black, with a broad transverse tawny yellow fascia about the middle. The borders of this band are irregular, and there is a narrow continuation of the tawny yellow along the suture to the scutellum.

The anterior coxal cavities are widely open behind and confluent. The long *legs* are black, except the coxae, the trochanters and the greater part of the femora which are tawny yellow. The front and middle coxae are large, exserted and conic, grooved on their outer sides and contiguous; the hind coxae are large and transverse, posteriorly grooved and contiguous. The trochanters of the middle and hind femora are conspicuously large, the femora are moderately thickened (incrassate). The tibiae have each two distal spurs, but those of the hind tibiae are lobe-like. The front and middle tarsi are five-segmented; the front tarsus is as long as the tibia; the first or basal segment is as long as the fifth, the other tarsal segments are progressively shorter; the first four segments are obconical and notched (bilobed), the fifth segment is the thinnest and it is club-shaped; the claws are simple, and each claw has an accessory lateral appendage which is claw-like but shorter and thinner. The tarsal segments are thickly fringed at the sides and beneath with short yellowish hairs. The hind tarsi are wanting in this type specimen.

The *abdomen* has five visible sterna. The first four sterna are about equal in length and yellowish red; the fifth, which is black, appears to be longer than the others and is narrowed, and its posterior border is arcuate. The surface of the abdominal sterna is closely punctulate, and the punctules bear moderately long recumbent yellowish hairs.

Length 9 mm.; breadth (across the middle of the elytra) 4½ mm.
Hab. Siberia (Oliv.).
See Plate 57.

Sub-family NEMOGNATHINAE

Tribe Nemognathini

95. *Euzonitis quadripunctata* (Fab.)

Coleopterorum Catalogus, pars 69 (F. Borchmann, 1917), Meloidae, etc. p. 154.

Spain, Italy, Turkey, Southern Russia.

SYN. *Mylabris 4-punctata* Fab., *Mant. Ins.* I, p. 217, No. 6 (1787); *Ent. Syst.* I, 2, p. 89, No. 10 (1792); *Syst. Eleuth.* II, p. 84, No. 15 (1801).
Mylabris imperialis Woll.
M. obliquata Motsch.
M. immaculata Escher.

As stated by Fabricius, the type of this species is in the Banks Collection, British Museum. The two Hunterian examples in Cabinet B, drawer 13, under label

'*Myl. 4-punctatus*
Fabr. MSS'

have been compared with the type and with modern examples in the British Museum, and they correspond.

Dr K. G. Blair[1] points out that the reference is incorrectly given in recent Catalogues as *Eleuth.* II, 1801, p. 84.

[1] 'Further Notes on the Fabrician Types of Heteromera (Coleoptera) in the Banks Collection' by K. G. Blair (*Annals and Magazine of Natural History*, Vol. V, Ninth Series, February 1920).

96. *Zonitis lineata* Champ.

Trans. Ent. Soc. London, 1896, p. 53. *Coleopterorum Catalogus*, pars 69 (F. Borchmann, 1917), Meloidae, p. 164.

Grenada.

SYN. *Lagria marginata* Fab., *Sp. Ins.* I, p. 159, No. 5 (1781); *Mant. Ins.* I, p. 93, No. 5 (1787).

Dryops marginata Fab., *Ent. Syst.* I, 2, p. 76, No. 6 (1792); *Syst. Eleuth.* II, p. 68, No. 6 (1801).

Oedemera marginata Oliv., *Ent.* III, 50, p. 8, pl. 1, fig. 7 (1795).

Zonitis strigata Wellm., *Deutsche Ent. Zeitschr.* 1910, p. 26.

The above synonyms are those given by Olivier, except *strigata* which is given by Borchmann.

The specimen under label

'*Lag. marginata*
Fabr. pag. 159, No. 5'

in Cabinet A, drawer 4, is in a very imperfect state, the head and the greater part of the left elytron and portions of the legs are wanting; it has been examined by Dr K. G. Blair and found to agree with *Zonitis lineata* Champ. As this insect answers the descriptions given by Fabricius and Olivier and corresponds with Olivier's illustration, it is apparently the type.

Description of Type, *Lagria marginata* Fab. The specimen is a *male*. Form elongate, the thorax small, much narrower than the elytra, which are long and nearly parallel-sided. The legs long. General coloration tawny brown, darker on the elytra between the reddish yellow borders and the median stripe and on the tibio-tarsal parts of the legs. The surfaces of the thorax, elytra and abdomen closely punctulate and rugulose, the puncturation being largely confluent and bearing very short and fine golden hairs.

Viewed from above, the *thorax* is nearly as long as broad; it is a little constricted at the base, rounded out about the middle and gradually narrowed to the front; its breadth is a little more than half that of the elytra at the base, and it is wedge-shaped in side view.

The *pronotum*, which is light tawny brown, is slightly excavate in front and straight across the base; the sides are bulged about the middle and are deeply inflexed, and the anterior and posterior angles are rounded. Upon the flattened disc, just beyond the middle, there are two small pits, one towards each side. The pronotum is marginate, finely so at the front, and its surface is irregularly and closely punctulate, especially between the middle and the base, where the puncturation is confluent with rugulose effect.

The *prosternum* is transverse and angulate between the bases of the front coxae. The mesopleura and metapleura are largely exposed, not covered in by the narrow side-inflexions of the elytra. The *mesepisterna* and *mesepimera* are long and the *mesosternum* is short; the oblong *metepisterna* are hollowed, and the *metasternum* is large and convex, with a blunt metasternal process extending between the hind coxae. The surfaces of the thoracic pleurites and sternites are finely rugulose with confluent puncturation and clothed with short and fine yellowish hairs.

The *scutellum*, light yellowish brown, is fairly large and triangular; the sides, being excavated towards the base, are angulate and the apex is rounded; the depressed surface is marked with a median longitudinal line and is finely rugulose, with confluent irregular punctulation bearing very short and fine yellowish hairs.

The *elytra*,[1] moderately and regularly convex, are long and narrow, longer than the body, with evenly rounded obtuse and incurved apices which enclose the end of the abdomen. The narrowly inflexed sides (viewed from above) are subparallel, slightly sinuate, being a little constricted behind the prominent shoulders. The outer and the sutural borders are finely marginate, but the margin becomes faint towards the apex. The elytral surface is closely punctulate and set with very short and fine golden hairs; the punctules are mostly confluent with rugulose effect. The coloration of the elytra is

[1] The left elytron is imperfect, only the basal portion remaining.

Lagria marginata Fab. ♂ × 10

dark tawny brown bordered with reddish yellow on the outer sides, round the apices, and along the suture; and on each elytron there is a narrow median longitudinal reddish yellow linear stripe, which is bent and which tapers off at each end, not reaching the base and the apex.

The *legs* are long and dark brown, except the femora which are light tawny with dark brown at the distal ends; and the front tibiae are tinged with blue. The front and middle coxae are long and roughly cylindrical, exserted and close together. The front coxae are slightly grooved; the middle coxae and the large transverse hind coxae are strongly grooved on the outer side for reception of the femora. All the tibiae have each two distal spines. The front and middle tarsi are five-segmented; the first or basal segment is the largest, the second is two-thirds the length of the first, the third and fourth are progressively shorter, and the fifth is nearly as long as the first. The hind tarsi are wanting in this type specimen (as noted by Olivier). The claws are simple and recurved; each claw has an inner comb of short denticles or spines and also a simple accessory bristle-like appendage of corresponding length.

The *abdomen* is light tawny coloured above; the second abdominal sternum is light tawny, and it is larger than the succeeding four which are dark brown. The surface of the abdomen is closely and irregularly punctulate and covered with fine recumbent light yellow hairs.

This type is evidently a male; the fifth abdominal sternum is broadly and deeply emarginate, and the middle portion of its posterior border has a large arcuate depression with a short deep slit at the top of the arch and a deep inner hollow which is perfectly smooth and nitid.

Length (body only) 12 mm.; breadth (across middle of elytra) 5½ mm.

Hab. South America (Fab. and Oliv.).

See Plate 58.

Family PYROCHROIDAE

97. *Neopyrochroa flabellata* (Fab.)

Coleopterorum Catalogus, pars 99 (K. G. Blair, 1928), Pyrochroidae, p. 3. *Catalogue of the Coleoptera of America, North of Mexico* (Charles W. Leng, 1920), p. 161.

North America.

SYN. *Pyrochroa flabellata* Fab., *Mant. Ins.* I, p. 162, No. 2 (1787); *Ent. Syst.* I, 2, p. 105, No. 3 (1792); *Ent. Syst.* Suppl. II, p. 105 (1798); *Syst. Eleuth.* II, p. 109, No. 3 (1801); Oliv. *Ent.* III, 53, p. 5, pl. 1, fig. 3 (1795).

The insect in Cabinet B, drawer 6, under label

'*Pyr. flabellata*
Fabr. MSS'

is the type; it answers the descriptions given by Fabricius and Olivier, and it has been compared with a modern example of this species in the British Museum Collection.

Olivier's figure of this type is a good representation, except that the subpectinate character of the antennae is not clearly shown.

Description of Type, *Pyrochroa flabellata* Fab. This example is a *male*. Form elongate and rather narrow; the head subtriangular and exserted, conspicuously projecting, horizontal and strongly constricted behind the emarginate eyes, and with subpectinate antennae; the thorax subquadrate but rounded at the angles, and broader than the head; the long elytra much wider than the thorax, parallel-sided, with very rounded apices divergent at the suture. The front and middle tarsi five-segmented, the hind tarsi four-segmented, and the penultimate tarsal segments bilobed. The head, the basal segments of the antennae, the thorax, legs and abdomen glossy light reddish yellow; the eyes, the antennae (except the proximal segments), and the elytra uniformly glossy brownish black; the surface of the head unevenly impressed and lightly coated with fine silky hairs.

The *head* is subtriangular and conspicuously exserted, projecting horizontally; it is strongly constricted at the neck and somewhat dilated behind each eye. The vertex is convex and narrowed between the eyes; the fronto-clypeus is strongly impressed on the frons, there is a deep pit on the inner side of each antennal insertion and the clypeal portion is narrowed and convex; the transverse clypeo-labral suture is almost straight, the labrum is transverse and rounded in front. The maxillary palps are light red (yellowish red), long and exserted, with the distal segment elongate and cleaver-shaped. The mandibles are brownish red. The mentum is long and broad and the adjoining post-genae are markedly dilated; at the junction of these parts with the large convex gula there are two conspicuous gular pits in front of the gular sutures, which are directed obliquely outwards.

The coloration of the head is reddish yellow and its uneven surface is punctulate, the punctules bearing moderately long and fine silky hairs.

The eyes are conspicuous, reniform (strongly emarginate) and black. The eleven-segmented antennae are inserted in front of and near the eyes, and are black, except the two first or proximal segments which are reddish yellow; the first antennal segment is long and clavate, the second segment is conical and short, less than half the length of the first segment and about half the length of the third, the third segment is serrate, and the remaining eight segments are subpectinate.

The glossy reddish yellow *prothorax*, which is coated with fine silky hairs, has a subquadrate appearance (upper view); but the sides are a little angulate beyond the middle, and the anterior and posterior angles are rounded. Vertically the sides are rounded and the *pronotum* is not clearly marked off from the proepisterna. The front of the prothorax is marginate round the narrow neck-like constriction of the head; the base of the pronotum is marginate, and within the base there is an intervening transverse furrow. The surface of the pronotum is very uneven and the disc forms a double convexity on each side of a median longitudinal impressed line.

The *prothoracic episterna* are large, unevenly convex and light red. The *metepisterna* are oblong, narrow, and tapering towards the small *metepimera*. The *prosternum* is long before the front coxae and the prosternal process is short, triangular and very thinly produced between the front coxae, which are almost contiguous. The *mesosternum* is large and convex and is produced as a process between the middle coxae. The *metasternum* is very long and unevenly convex; its posterior portion is divided by a narrow but deep median sulcus. The surface of the sternal parts is imbricate and thinly covered with fine silky hairs.

The small reddish yellow *scutellum* is narrowed and rounded posteriorly; it is slightly convex and fringed with fine silky hair.

The *elytra* are, across the base, fully twice the breadth of the base of the thorax; they are long, rather narrow, roughly parallel-sided, with very rounded apices divergent at the suture, and finely marginate only from the shoulder to about the middle of their outer sides where the margination becomes obsolete; the apical portions of the elytra and the parts before the middle are distinctly wider than the basal portions, consequently the outer sides are a little sinuate; the bases are sinuate. The elytra are convex, markedly so at the shoulders, which are prominent and rounded, and over the apical portions; the elytral surface is closely tuberculate with rugulose effect, the fine tubercles bearing punctules with fine black hairs; and it is also closely striate with wavy sub-parallel longitudinal lines extending from the base towards the apex. The coloration of the elytra is uniform glossy brownish black. The epipleura are very narrow and short, ending with the margin near the middle of each outer side.

The anterior coxal cavities are widely open behind and confluent. The *legs* are reddish yellow and long; the front

Pyrochroa flabellata Fab. ♂ × 6

and middle coxae are long and subcylindrical, exserted and contiguous; the hind coxae are transverse, elongate and almost contiguous; the antecoxal pieces are rib-like. The front and middle tarsi are five-segmented, the hind tarsi are four-segmented; the first tarsal segment is elongate, the penultimate tarsal segment is bilobed with the lobes dilated, the distal segment is as long as the first but thinner, and the claws are simple.

The *abdomen* is glossy reddish yellow and closely punctulate with fine silky hairs. There are six abdominal segments; the first five sterna are about equal in length, but the posterior border of the fifth is arcuate; the sixth sternum is shorter, greatly narrowed and tapered, and its posterior border is arcuate.

Length 15 mm.; breadth (across the middle of the elytra) 5 mm.
Hab. America (Fab. and Oliv.).
See Plate 59.

SUPPLEMENT TO PART I

Revised references according to the *Coleopterorum Catalogus* (Junk and Schenkling), and notes on certain types.

Page 11 5. ***Procerus scabrosus*** (Oliv.)

Coleopterorum Catalogus, pars 91 (E. Csiki, 1927), Carabinae, I, p. 35. *Catalogus Coleopterorum* (Gemminger and Harold, 1868), I, Carabidae, p. 56.

Turkey.

SYN. *Carabus scabrosus* Oliv., *Ent.* III, 35, p. 17, pl. 7, fig. 83 (1795).

Page 13 6. ***Scaphinotus elevatus*** (Fab.)

Coleopterorum Catalogus, pars 92 (E. Csiki, 1927), Carabinae, II, pp. 317, 318. *Catalogus Coleopterorum* (Gemminger and Harold, 1868), I, Carabidae, p. 84.

Florida.

SYN. as given on p. 13 with the addition of *Cychrus elevatus* Fab., *Syst. Eleuth.* I, pp. 166, 167, No. 5 (1801).

Page 15 7. ***Scaphinotus heros*** Harris

As pointed out by Professor E. C. van Dyke, and according to his recent revision of the Genus *Scaphinotus*, the figure (Plate 6) would appear to be that of var. *heros* Harris whereas the description agrees with the true *unicolor* Fab. Re-examination of the type shows that in ordinary daylight there is no trace of violet; but in strong light a faint suffusion of violet is perceptible and this, unfortunately, was intensified in the process of producing the plate.

The coloration, not being decidedly violaceous, is that of *unicolor* Fab. The margin or bead of the pronotum is thicker than the outer antennal segments; its thickest part (behind the middle) is as thick as the first or basal segment of the antenna. In that respect the type agrees with *heros* Harris.

The elytral raised lines (flattened costae), those from the suture to near the outer border, are distinctly wider than the striate-punctate interstices, which is a feature indicative of *unicolor* Fab. The Hunterian insect would appear to be the type of *Scaphinotus unicolor* (Fab.). Accordingly the name and references should read as follows:

Scaphinotus unicolor (Fab.)

Coleopterorum Catalogus, pars 92 (E. Csiki, 1927), Carabinae, II, pp. 318, 319. *Catalogue of the Coleoptera of America, North of Mexico* (Charles W. Leng, 1920), p. 43.

U.S.A. (Alabama, Southern States).

SYN. *Carabus unicolor* Fab., *Mant. Ins.* I, p. 198, No. 38 (1787); Oliv. *Ent.* III, 35, p. 47, pl. 6, fig. 62 (1795).
var. *heros* Harris, *Boston Journ. Nat. Hist.* II, 1839, p. 196.
var. *shoemakeri* Leng.

To the description of the *elytra* (on p. 17) the following addition should be made: There is a slight yet distinct sinuation, just behind the middle of each outer side.

Page 17 8. *Leistus ferrugineus* (Linn.)

Coleopterorum Catalogus, pars 92 (E. Csiki, 1927), Carabinae, II, p. 351. *Catalogus Coleopterorum* (Gemminger and Harold, 1868), I, Carabidae, p. 54.

North and Middle Europe.

SYN. *Carabus ferrugineus* Linn., *Syst. Nat.* ed. X, 1758, p. 415; *Fauna Suec.* ed. II, 1761, p. 221.
C. spinilabris Panz., *Fauna Germ.* 39, 1797, nr. 11; Fab., *Syst. Eleuth.* I, p. 204, No. 189 (1801).
C. rufescens Clairv.

Page 20 9. *Pasimachus depressus* (Fab.)

Coleopterorum Catalogus, pars 92 (E. Csiki, 1927), Carabinae, II, p. 445. *Catalogue of the Coleoptera of America, North of Mexico* (Charles W. Leng, 1920), p. 47. *Catalogus Coleopterorum* (Gemminger and Harold, 1868), I, Carabidae, p. 176.

U.S.A. (New York, Florida, New Jersey, North Carolina, Pennsylvania).

SYN. as given on page 20 with the addition of *Scarites complanatus* Gmel.; *Pasimachus laevis* J. Lec.; *P. champlaini* Casey; *P. carolinensis* Casey; var. *morio* (South Carolina) J. Lec.

Page 22 10. ***Pasimachus marginatus*** (Fab.)

Coleopterorum Catalogus, pars 92 (E. Csiki, 1927), Carabinae, II, p. 446. *Catalogue of the Coleoptera of America, North of Mexico* (Charles W. Leng, 1920), p. 47. *Catalogus Coleopterorum* (Gemminger and Harold, 1868), I, Carabidae, p. 177.

Florida, South Carolina, Louisiana, Texas.

SYN. as given on p. 22 with the addition of var. *crassus* (North Carolina) Casey.

Page 25 11. ***Scarites subterraneus*** Fab.

Coleopterorum Catalogus, pars 92 (E. Csiki, 1927), Carabinae, II, p. 487. *Catalogue of the Coleoptera of America, North of Mexico* (Charles W. Leng, 1920), p. 47. *Catalogus Coleopterorum* (Gemminger and Harold, 1868), I, Carabidae, p. 187.

U.S.A. (California, Florida, Louisiana, Indiana).

SYN. as given on p. 25.

Page 26 For 12. ***Lebia turcica*** (Fab.) read
12. ***Lebia scapularis*** Fourcroy

Coleopterorum Catalogus, pars 124 (E. Csiki, 1932), Harpalinae, VII, p. 1323. *Catalogus Coleopterorum* (Gemminger and Harold, 1868), I, Carabidae, p. 141.

Middle Europe, Caucasus.

SYN. as given on p. 26.

Page 27 13. ***Metabletus truncatellus*** (Linn.)

Coleopterorum Catalogus, pars 124 (E. Csiki, 1932), Harpalinae, VII, p. 1419. *Catalogus Coleopterorum* (Gemminger and Harold, 1868), I, Carabidae, p. 133.

Europe, Siberia.

SYN. as given on p. 27.

Page 28 14. *Callida ruficollis* (Fab.)

Coleopterorum Catalogus, pars 124 (E. Csiki, 1932), Harpalinae, VII, p. 1442. *Catalogus Coleopterorum* (Gemminger and Harold, 1868), I, Carabidae, p. 116.

Sierra Leone, Natal.

SYN. as given on p. 28.

Page 30 15. *Pheropsophus aequinoctialis* (Linn.)

Coleopterorum Catalogus, pars 126 (E. Csiki, 1933), Harpalinae, VIII, pp. 1604, 1605. *Catalogus Coleopterorum* (Gemminger and Harold, 1868), I, Carabidae, p. 102.

Tropical America (Brazil, etc.).

SYN. as given on p. 30 with the addition of *Cicindela aequinoctialis* Linn., *Amoen. Acad.* VI, p. 395 (1763).

Pages 33–37 16. *Agonoderus pallipes* (Fab.)

Two specimens under label *Carabus pallipes*. It was a difficult matter to examine these properly for description, owing to their frail state; and for the same reason they could not safely be removed for illustration. In the description of the larger one, which was chosen provisionally as the possible type, the pronotum is described as 'orbicular'; but, as suggested by Dr Fall, it is rotundate rather than orbicular. The reference to the marginal setae of the pronotum (p. 34 and lines 11, 12 and 13 from bottom) should be emended to read 'on each side margin there are two long setae (one about the middle and one at the hind angle)'. The presence of a distinct mentum tooth should be included in the description of the head.

According to Dr Fall, the following characters differentiate this insect from any known *Agonoderus*: The rotundate pronotum with two marginal setae; the elytra strongly striate-punctate and the scutellar striole situated not between the first and second striae (as in *Agonoderus*) but between the suture and the first stria, and also three large punctures on the third interspace; two supraorbital setae and a distinct mentum tooth, whereas in *Agonoderus* there is only one such seta and the mentum tooth is absent. Therefore the larger specimen is clearly not *Agonoderus pallipes*, but probably it is *Agonum limbatum* (Say) formerly *Platynus limbatus* (Say), *Coleopterorum Catalogus*, pars 115 (E. Csiki, 1931), Harpalinae, v, p. 843.

The smaller specimen under label *Carabus pallipes* is an *Agonoderus* but apparently it is not *pallipes* Fab.; it corresponds closely with modern examples of *Agonoderus comma*, submitted by Dr Fall, and probably it is the Fabrician type of *Carabus comma* acquired by Hunter from the Drury Collection. Accordingly the name and references should read as follows:

Agonoderus comma (Fab.)

Coleopterorum Catalogus, pars 121 (E. Csiki, 1932), Harpalinae, VI, p. 1188. Casey, *Mem. Col.* v, 1914, p. 295.

Virginia, Lake Superior.

SYN. *Carabus comma* Fab., *Syst. Ent.* p. 248, No. 63 (1775); *Sp. Ins.* I, p. 312, No. 78 (1781); *Mant. Ins.* I, p. 205, No. 112 (1787); *Ent. Syst.* I, I, p. 165, No. 179 (1792); *Syst. Eleuth.* I, p. 207, No. 201 (1801).

Carabus pallipes Say nec. Fab., *Trans. Amer. Philos. Soc.* II, 1823, p. 38; *Complete Writings*, II, 1859, p. 465.

Agonoderus dorsalis J. Lec.

A. similis J. Lec.

In the description given, at line 8 from bottom of p. 36, for 'The scutellary striole unites with stria 1' read 'The scutellary striole is rather long and it unites with stria 1'.

Page 48 22. *Copris fricator* (Fab.)

In the *Fauna of British India, Coleoptera Lamellicornia*, Part III (Coprinae), p. 106, Mr G. J. Arrow writes: 'In the Junk Catalogue Dr Gillet has called this species (*Copris indicus*) *Copris fricator* (Fab.), but the specimen in the Glasgow University Museum believed to be the type of that species is the North American insect known as *C. anaglypticus*'.

Fabricius was often wrong in his localities and so 'Eastern India' might well have been an error. I have compared this Hunterian *Copris fricator* (Fab.) with modern examples of *C. anaglypticus* in the Bishop Collection and it apparently agrees, except that the cephalic horn is quite different and there is no groove on the metasternum. It answers the description of *Copris indicus* (given by Mr Arrow) with the exception of the following points: The cephalic horn is *flattened behind*, the front angles of the pronotum are sharp (? 'broadly truncate'), the pygidium is *rather strongly* punctured, the metasternum is *without a median groove*, and the terminal spur of the front tibia has the tip strongly bent, hook-like. The length of this insect is more than 15·5 mm., the limit stated by Mr Arrow.

To the description the following additions should be made: Line 14 from top of p. 49, after the word 'black' insert 'The mouth-parts and the hairs of the legs red'. At end of line 4 from top of p. 50 insert 'The metasternum is lightly punctured in front and the puncturation becomes obsolete about the middle'. Line 21 from top of p. 50, after the word 'spine' insert 'which has the tip strongly bent, hook-like'.

Page 52 24. *Circellium bacchus* (Fab.)

Insert the following reference:

Coleopterorum Catalogus, pars 38 (J. J. E. Gillet, 1911), Coprinae, I, p. 15.

Page 53 25. *Phanaeus bellicosus* (Oliv.)

Insert the following reference:

Coleopterorum Catalogus, pars 38 (J. J. E. Gillet, 1911), Coprinae, I, p. 81.

For 'Brazil' read 'Guiana, Brazil'.

Page 57 27. ***Onthophagus pactolus*** (Fab.)

Coleopterorum Catalogus, pars 90 (A. Boucomont and J. J. E. Gillet, 1927), Coprinae, II, p. 143. *Catalogus Coleopterorum* (Gemminger and Harold, 1869), IV, Scarabaeidae, p. 1034.

India.

SYN. as given on p. 57.

Page 75 36. ***Enema infundibulum*** Burm. (=***Enema pan*** Fab. ♂ var.)

See *Coleopterorum Catalogus*, pars 156 (G. J. Arrow, 1937), Scarabaeidae: Dynastinae, p. 65.

Page 77 37. ***Enema pan*** (Fab.)

See the above reference.

Page 80 38. ***Strategus titanus*** (Fab.)

See *Coleopterorum Catalogus*, pars 156 (G. J. Arrow, 1937), Scarabaeidae: Dynastinae, p. 74 (*Strategus aenobarbus* Fab.) and p. 76 (*Strategus simson* L.).

Page 82 ***Scarabaeus eurytus*** Fab.

See the above reference.

Page 83 39. ***Golofa hastata*** (Fab.)

Insert the following reference:

Coleopterorum Catalogus, pars 156 (G. J. Arrow, 1937), Scarabaeidae: Dynastinae, p. 100.

Page 85

40. ***Dynastes (Xylotrupes) gideon*** var. ***oromedon*** Drury

Insert the following reference:

Coleopterorum Catalogus, pars 156 (G. J. Arrow, 1937), Scarabaeidae: Dynastinae, p. 96.

Page 100 51. ***Necrodes surinamensis*** (Fab.)

Insert the following reference:

Coleopterorum Catalogus, pars 95 (M. H. Hatch, 1928), Silphidae, II, p. 123.

For 'North America' read 'North America, Northern South America (?)'.

INDEX

The original names are printed in *bold* type.

Printed in the United States
By Bookmasters